ECOLOGY

by TAYLOR R. ALEXANDER
and
GEORGE S. FICHTER

Illustrated by
RAYMOND PERLMAN

Under the general editorship of
VERA R. WEBSTER

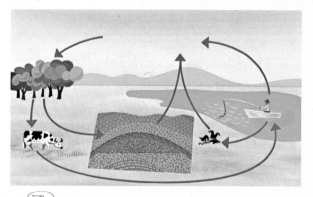

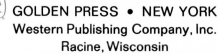
GOLDEN PRESS • NEW YORK
Western Publishing Company, Inc.
Racine, Wisconsin

FOREWORD

In recent decades, man has become a disturbing factor in the natural, complex system of checks and balances between earth's living and nonliving components. Locally and even on a global basis, examples of man's disruptive activities are numerous. The recognition of man's capacity to upset natural processes has given rise to the great current interest in ecology.

Man's cause-and-effect role in ecological crises is well documented in numerous publications, hence this aspect of ecology is not dealt with at length in this book. Rather, this book gives a general approach to ecology to demonstrate how man and all other living things relate to their environment. Only by understanding these basic principles can man again achieve and then maintain the sort of environmental quality that was enjoyed prior to the mid-1900's.　　　　　　**T.R.A. and G.S.F.**

PHOTO CREDITS: p. 4, NASA; p. 19, T. R. Alexander; p. 33, T. R. Alexander; p. 38, Leon Follmer, Illinois State Geological Survey; P. 48, NASA; p. 58, Forest Service, USDA; p. 63, T. R. Alexander; p. 65, Alan Crook; p. 66, North Carolina Division of Commercial and Sport Fisheries; p. 71, Herbert C. Jackson; p. 73, Florida Department of Natural Resources; p. 74, T. R. Alexander; p. 85, U.S Soil Conservation Service; p. 87, T. R. Alexander; p. 96, T. R. Alexander; p. 100, Concentration Regional Analysis, University of Wisconsin—Green Bay; p. 101 (top), Kentucky Department of Natural Resources; p. 101 (bottom), U.S. Corps of Engineers; p. 103, 104, 105, U.S. Soil Conservation Service; p. 106, Ray Perlman; p. 108, Forest Service, USDA; p. 109, Michigan State University Agricultural Experiment Station; p. 118, Ray Perlman; p. 127, Ray Perlman; p. 128, NASA; p. 130, Forest Service, USDA; p. 131 (top), ASCS-USDA; p. 131 (bottom), reprinted by permission from copyrighted Kodak publication; p. 132, Westinghouse; p. 133, Laboratory for Application of Remote Sensing, Purdue University; p. 140, Ray Perlman; p. 145, David B. Marshall, U.S. Sport Fisheries and Wildlife; p. 150, Alan Crook; p. 151, Washington Department of Fisheries.

CONTENTS

Earth photographed from Apollo 11 on its way to the moon.

ECOLOGY

Ecology is a subdivision of the science of biology that deals with the dependency and interaction between the earth's living (biotic) and nonliving (abiotic) systems.

Earth can be described as a system of reacting elements and their chemical compounds, the energy being supplied by the sun. Living things are involved in many of these chemical reactions. All of the elements, compounds, and organisms are components of the biosphere —that part of the earth where life exists. Ecology is the science that interprets the biosphere.

Another definition stresses the origin of the word ecology, which comes from the Greek word *oikos*, meaning house. By this interpretation, ecology becomes the study of the environment and the ways that living things "house" themselves in it.

Ecology can also be defined as the study of ecosystems, which are functional units resulting from the interaction between plants and animals and the physical and chemical components of their environment. The largest conceivable ecosystem is the entire earth. The general concept of an ecosystem is a smaller, self-perpetuating level of organization, interlocking with similar systems to form the biosphere. Though there are numerous such systems on earth and the interaction between living things and their environments are complex, the earth's total ecosystem tends to be stable. Ecologists refer to this steady, self-regulating state as homeostasis.

Earth's life-inhabited regions related to other spheres.

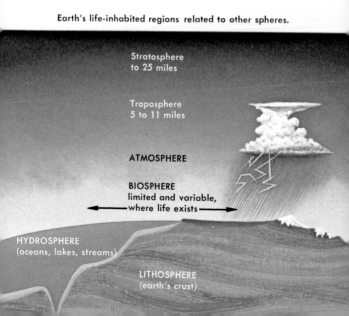

Stratosphere
to 25 miles

Troposphere
5 to 11 miles

ATMOSPHERE

BIOSPHERE
limited and variable,
where life exists

HYDROSPHERE
(oceans, lakes, streams)

LITHOSPHERE
(earth's crust)

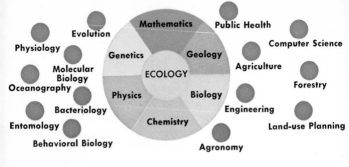

ECOLOGISTS began to specialize early. As a result, their viewpoints became too narrow for solving the broad problems brought about by the impact of increasing human population and industrialization on the total world environment.

The most useful approach for solving the complex environmental problems of today is by ecological teams. This method puts to use the knowledge and skills from such other biological specialties as taxonomy, genetics, physiology, and evolution. Chemistry, physics, geology, agronomy, mathematics, computer science, and government specialists are also utilized. This dynamic viewpoint of modern ecology recognizes that the makeup of communities changes with time. Their past can be reconstructed and their future predicted.

The study of unaltered environments is no longer possible, for no part of the earth's surface is pristine. Today's ecology is largely applied. Plants, animals, and microorganisms are manipulated in their environments to achieve a desired result. Wildlife management, forestry, soil conservation, outdoor recreation, human ecology—all of these are among the many special fields of applied ecology.

AUTECOLOGY deals with individual species rather than with groups of species or with communities. Single environmental factors may be worked with also to learn how they influence species. Because complex equipment and facilities are often needed, the research is commonly done in a laboratory rather than in the field. This early approach to ecology is not easily distinguished from plant or animal physiology.

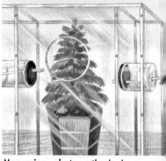
Measuring photosynthesis in a growth chamber in laboratory.

SYNECOLOGY is concerned with problems solved by studies of natural groups or communities, or it may involve study of the many factors affecting communities. Emphasis, however, is on the group rather than on a single species or on one factor as in autecology. Most of the current work is with synecology, analyzing the effect of the total environment on complex communities in their natural setting. Modern methods also make use of computers and other means of collecting and interpreting quantities of data (p. 122).

SPECIAL KINDS OF ECOLOGY are aquatic, marine, freshwater, desert, forest, island, and others pertaining to specific geographical areas. Other ecologists concentrate on particular groups of organisms or life conditions. A paleoecologist studies fossils to learn about communities and environments of the past. All of these specialties are governed by the same basic concepts, and each demands information from the others. Ecology's complexity today simply reflects the better understanding of the interaction of all processes on earth.

A marine ecologist and a paleoecologist—two specialists at work.

Ecosystems are varied, intricately interwoven assemblages of plants and animals. In the illustration here, a relationship is shown between freshwater, estuarine, and ocean ecosystems. Similarly, a forest blends with a saltwater marsh that, in turn, is closely tied to the aquatic ecosystems. Each is distinct but nevertheless dependent on the other.

ECOSYSTEMS, regardless of size, consist of an assemblage of plants and animals linked by a fundamental need: food. Food must be in constant supply for growth and maintenance of the body and also to provide the work energy needed by the living system. All of the energy and materials originate basically in the physical environment. In the ecosystem, living things absorb, transform, and circulate energy and matter and then release it to the environment again. The final release comes at death.

Some of the transport, use, and temporary storage of energy and materials comes about by such biological processes as photosynthesis, respiration, absorption, digestion, transpiration, and excretion. The constant interactions of plants and animals in their ecosystem are the pathways by which matter and energy are distributed. A food chain (p. 12) is a pathway of this sort.

Energy and materials are also stored and circulated by physical processes in the nonliving or abiotic part of the ecosystem. Precipitation, evaporation, erosion, deposition, and a variety of chemical reactions are among the processes that are constantly changing the availability of energy and matter in the system.

The kinds and numbers of organisms forming an ecosystem also affect the movement of energy and matter through the system. Regulatory mechanisms, both within the organisms and also in the abiotic portion of the ecosystems, provide a constant feedback on the matter and energy of which the ecosystems are composed. Growth and reproduction, mortality, immigration and emigration. These are among the numerous and important balancing mechanisms that affect ecosystems. They influence also behavior, physiology, and population numbers.

All ecosystems are "open," which means that energy and matter continually escape from the ecosystem as they are used by living organisms. Unless they are replaced, the ecosystem will collapse. Usually the replacement comes from the abiotic environment, but sometimes it occurs between adjacent ecosystems. Often the pathway is difficult to determine.

The interlacing of smaller ecosystems by energy and matter is a major factor favoring consideration of the earth's biosphere as a single ecosystem. All changes in an ecosystem are echoed throughout the biosphere in varying degrees of intensity.

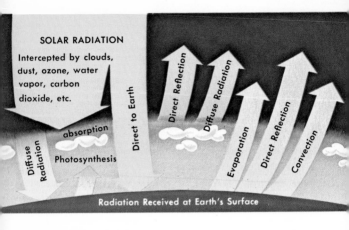

SOLAR RADIATION
Intercepted by clouds, dust, ozone, water vapor, carbon dioxide, etc.

Diffuse Radiation

absorption

Photosynthesis

Direct to Earth

Direct Reflection

Diffuse Radiation

Evaporation

Direct Reflection

Convection

Radiation Received at Earth's Surface

ENERGY received as solar radiation drives all life and meteorological processes on earth. Without this energy, on which the production of basic food by photosynthesis depends, ecosystems would cease to function, for food chains would be destroyed at their first level (p. 14).

Solar energy affects ecosystems in many other ways. For example, energy exchanges in the physical environment control the weather. Other energy-exchange systems are important to the survival and comfort of living things. These include such things as warming from the sun. Ecologists must therefore consider all aspects of an ecosystem's energy budget.

SOLAR FLUX is the total radiant energy of all wave lengths reaching the earth. Its amount varies in different seasons, locations, and times of day. Only about half of the energy that enters the troposphere reaches the earth's surface, and of this amount, less than 2 percent is used in photosynthesis. A much larger percentage of the sun's energy is involved in the earth's climate and in other physical factors. Ecologists must know how this energy is used and exchanged in the biosphere.

ENERGY TRANSFERS between organisms and their environment are often complex. Special techniques and equipment are needed for their analysis and measurement. Radiation intensity, relative humidity, wind, and air temperature are among the kinds of data needed.

Each kind of organism has a different capacity for absorbing and releasing radiation. Color is an important factor. Black or dark colors absorb radiation much more readily than do lighter colors, hence darker surfaces tend to be hotter.

The amount of area exposed is also important in radiation absorption. To get more heat, some animals flatten themselves; others expose more surface by extending appendages. Plants react to variations in light with

adjustments by well-known phototrophic responses.

An energy exchange occurs, too, when an organism is in physical contact with a surface. If the surface is cooler, heat is lost by conduction. If the surface is warmer, heat passes to the organism's body.

Organisms are also coupled to their environments by convection, the energy exchange taking place in the boundary layer of water or air surrounding their bodies. This differs on a still day as compared to a windy day or may be varied by the movement of water. For example, a mildly cool day based on the temperature may seem very cold if there is a wind. A day when the temperature is much lower may not "feel" as cold if there is no wind blowing.

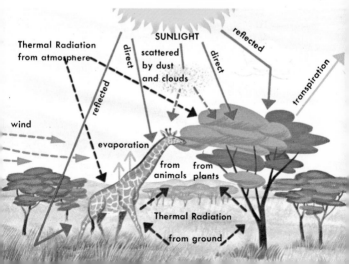

Thermal Radiation from atmosphere

SUNLIGHT

reflected

direct

scattered by dust and clouds

direct

transpiration

wind

reflected

evaporation

from animals

from plants

Thermal Radiation from ground

FOOD CHAINS

Food chains are the pathways by which living things obtain, use, and transfer energy. Plants trap solar energy in their manufacture of food by photosynthesis. They are the basic food producers. Animals are the ultimate consumers. Animals that feed exclusively on

GRAZING FOOD CHAINS depend directly on the input of solar energy to produce the basic food of the ecosystem. They are essentially systems in which living organisms serve as food for other living organisms that occupy a higher level in the food chain. At the end of a chain, for example, man may eat a fish that had lived on smaller fish. The smaller fish, in turn, had fed on still smaller animals that were the grazers in the chain, feeding on the phytoplankton or other green plants that trapped the sun's energy and produced the food which started the chain.

Very short grazing chains are common in agriculture where grain is grown to feed an animal that man eats. Since potential energy is lost at each food transfer, the length of the chain controls the number of consumers at each level. Many more people could be fed if the grain were eaten directly.

In natural ecosystems, food chains tend to be long and complicated, interwoven with others. Only a few carnivores occupy the top position, with an increase in organisms at each lower level.

DETRITUS FOOD CHAINS are very important in the energy flow of an ecosystem. Their base is the unused organic material that is produced in the grazing food chains. They depend also on the sun's energy, though indirectly.

As plants and animals die, their bodies decay, and their chemical components are returned to the ecosystem cycle. Large amounts of organic materials are also voided as excrement or released by living organisms. As examples, tree limbs and leaves make up the decaying litter on the forest floor, and dead seaweeds and marine animals accumulate along shores and on the sea floor. This constant release of organic wastes, which occurs in every ecosystem, provides food for the many different kinds and numerous small organisms (called reducers) and for detritus-feeding organisms (called detritivores). These may themselves become food for predators.

In most ecosystems, the detritus chain is closely interlocked with the grazing food chain. Both contribute to the total energy flow and the mineral recycling within the system.

plants are herbivores. Those feeding only on animals are carnivores. Omnivores are animals that feed on both plants and animals. In all cases, however, plants are the basic source of food.

In grazing food chains, green plants serve directly as food for animals that, in turn, are food for other animals. In detritus food chains, nonliving organic matter is the food and energy source. Food chains usually overlap, their interwoven complex making up the food web of the ecosystem.

Shown below is the relationship between the two basic food chains.

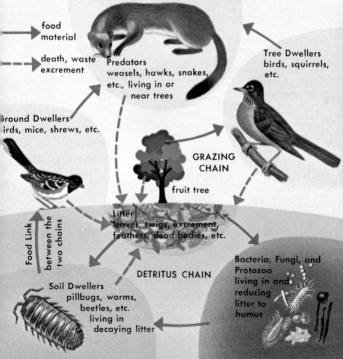

TROPHIC LEVELS in food chains follow the laws of thermodynamics: (1) every system's energy inflow must be balanced by an outflow; and (2) with each energy transfer, energy is also lost as unusable heat. The number of links in a food chain is thus limited by the amount of energy available to each succeeding link.

FIVE TROPHIC LEVELS shown below, occur in most complex natural communities:

1. **producer level**, consisting of photosynthetic and chemosynthetic plants.

2. **herbivore level**, made up of primary consumers or plant-eating animals.

3. **first carnivore** level, comprised of animals that feed as predators, scavengers, or parasites. These are the secondary consumers.

4. **secondary carnivore** level, consisting of animals that feed on animals of the first carnivore level. These are the tertiary consumers.

5. **tertiary carnivore** level, made up of animals that feed on the secondary carnivores.

Omnivores and carnivores may feed on members of more than one trophic level. This often makes it difficult to assign an animal to only one level. Man's diet consists of a mixture of food derived from both plants and animals, hence he derives his food at all four consumer levels in food chains.

Tertiary Carnivore Level

Secondary Carnivore Level

First Carnivore Level

Herbivore Level

Producer Level

Water or Soil

sunlight, water, minerals, CO_2

BIOTIC

ABIOTIC

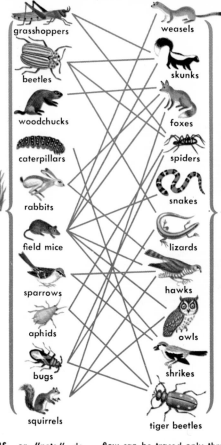

grasshoppers

weasels

beetles

skunks

woodchucks

foxes

caterpillars

spiders

rabbits

snakes

field mice

lizards

sparrows

hawks

aphids

owls

bugs

shrikes

squirrels

tiger beetles

FOOD WEBS, or "nets," describe the interlocking of food chains in the ecosystem. The great diversity of producers and consumers allows animals to select from a broad range of food sources. In complex communities, the food and energy flow can be traced only through these weblike interconnections.

In simple, man-made systems, such as a farm pond (p. 149) or a corn field, the flows of both food and energy can be explained or understood in terms of the basic food chain.

15

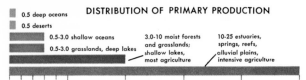

DISTRIBUTION OF PRIMARY PRODUCTION

0.5 deep oceans

0.5 deserts

0.5-3.0 shallow oceans

0.5-3.0 grasslands, deep lakes

3.0-10 moist forests and grasslands; shallow lakes, most agriculture

10-25 estuaries, springs, reefs, alluvial plains, intensive agriculture

Grams of dry organic matter per square meter per day

ECOLOGICAL PYRAMIDS illustrate the trophic levels in a food chain or an ecosystem. Large numbers of organisms, materials, and energy are necessary for the first trophic level, which forms the base of the pyramid. Each successive level involves less, shaping the pyramid.

The three kinds of pyramids commonly constructed are: productivity, numbers, and biomass. Their shapes will vary depending on when the data is taken. In spring, because of the greater abundance of producer organisms at that time, the base of the pyramid may be much broader than in winter.

PRODUCTIVITY PYRAMIDS show the amount of food available at each trophic level. The first trophic level, called the net primary production, consists of the quantity of food (organic matter) produced minus the amount used by the producer plants for their growth, repair, and respiratory energy.

Herbivores, which constitute the second level, depend on the net primary production for their food supply. It is possible for herbivores to eat so many of the food producers that the ecosystem collapses, but their population is usually regulated—typically by the carnivores of the next trophic level. A strong survival principle applies at all trophic levels.

Only a portion of the food eaten by herbivores and carnivores is retained in their bodies. Some parts of the food are not digestible and are cast out. Much is used for immediate energy in respiration. Ecologists generally agree that there is an average efficiency of about 10 percent—that is, each trophic level actually utilizes only about 10 percent of the production of the previous level.

For constructing productivity pyramids, the data used are usually the dry weight of food produced by the organisms at each level in a specified area and limited period of time. For an energy approach, the organic matter (food) is measured in calories.

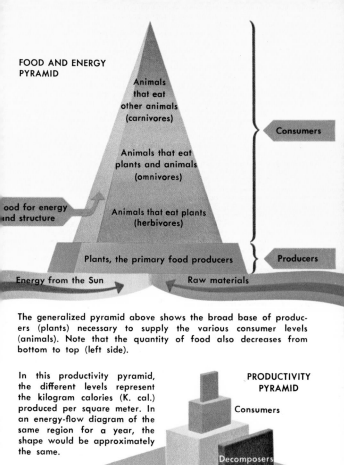

FOOD AND ENERGY PYRAMID

Animals that eat other animals (carnivores)

Animals that eat plants and animals (omnivores)

Animals that eat plants (herbivores)

Food for energy and structure

Plants, the primary food producers

Energy from the Sun

Raw materials

Consumers

Producers

The generalized pyramid above shows the broad base of producers (plants) necessary to supply the various consumer levels (animals). Note that the quantity of food also decreases from bottom to top (left side).

In this productivity pyramid, the different levels represent the kilogram calories (K. cal.) produced per square meter. In an energy-flow diagram of the same region for a year, the shape would be approximately the same.

PRODUCTIVITY PYRAMID

Consumers

Decomposers

Producers

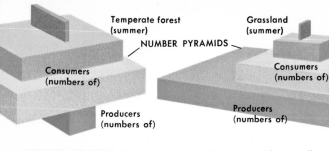

NUMBER PYRAMIDS — Temperate forest (summer), Grassland (summer), Consumers (numbers of), Producers (numbers of)

NUMBER PYRAMIDS show the population of the organisms that form each trophic level. In a typical pyramid, the number of individuals decreases at each higher level, for many producers are required to support fewer herbivores that, in turn, support still fewer carnivores.

The shapes of number pyramids vary with the kind of ecosystem. If the producers are extremely small, as in the case of phytoplankton, enormous numbers are necessary to feed the herbivore level. In contrast, a pyramid in which large trees are the producers has a small base. If insects or similar small animals feed on these trees at the herbivore level, then the pyramid will be reversed in shape between these two levels.

Number pyramids usually show the numbers of organisms in a specified area—as in a liter or quart, in a square meter, or in an acre or hectare.

BIOMASS PYRAMIDS, like the number pyramids, vary in shape in different kinds of ecosystems. The biomass is the total weight of all the organisms at each trophic level (p. 19). The weight generally decreases at each higher level of the pyramid. If the organisms of all the trophic levels are somewhat alike in size, the pyramid slopes gradually. If the producer organisms are very small, the pyramid will be inverted. The biomass of an area at a specified time is generally called the standing crop.

BIOMASS PYRAMIDS

Seasonal Changes in a Lake Pyramid — summer — winter — Abandoned Field — Consumers (dry weight) — Producers (dry weight)

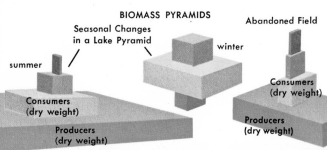

BIOMASS is the weight of all the producers, consumers, and reducers that exist in an ecosystem at a particular time. Biomass of live and dead material may be calculated separately, depending on kind of data needed.

WEIGHTS given for biomass may be measurements of fresh (live) or wet materials or of dry materials. For plants, dry weight is generally preferred because the amount of water that plants contain varies from time to time. The weights are usually given for specific areas of land or for volumes of water, such as square meters, acres, hectares, quarts, liters, etc.

Yield, such as tons of hay per acre, etc., is another way of expressing biomass. Standing crop and population density are also terms for biomass.

TO CALCULATE BIOMASS in a forest, all of the roots, stems, and leaves would have to be collected over a specified area, dried, and then weighed. To get an energy equivalent for biomass, the samples of plant or animal materials are dried, weighed and burned in a bomb calorimeter, results expressed as kilogram calories (K cal.) per unit of dry weight.

Daily food requirement of animals can be expressed in kilogram calories, and the availability of food at different trophic levels can be compared.

Ecologist is making a fresh or wet weight measurement of sawgrass on a research plot in the Florida Everglades.

Starlings normally interact negatively with man, but in plague outbreaks of some insects, they have been beneficial in bringing about control.

SPECIES INTERACTIONS

The relationships between living things in an ecosystem range from complete cooperation and dependency to total antagonism and competition. Between these extremes is a more or less neutral condition in which the effect of one organism on another is not apparent or is indirect. Each organism, though, is a part of the environment of all other living things and has at least some effect on it; each also influences to some degree the physical environment in which it lives.

Interactions vary from season to season and from one life stage to another as the needs and sometimes also the structure and size of the organism and the ecosystem change. All interactions have a bearing both on the survival of the species and also on the environmental niche occupied by the species. Sometimes, for example, a population is actually helped when it is preyed upon, even though many individuals are killed in the process. The victims are generally the less fit and the older individuals that have passed maturity and have lost the ability to produce more young.

PLANT-TO-PLANT INTERACTIONS occur between different species and also between individuals of the same species. Easily observed effects are the competition for water, the influence of shading, and the spread of diseases. Other interactions are not as easily determined.

SHADE produced by plants may be beneficial or harmful to other plants, depending on the species. Some plants require full exposure to sunshine. Many grasses are of this type.

Some species of trees require shade. These species are generally shorter than those that produce shade for them. If the taller trees are cut or destroyed by disease, those demanding shade will in time disappear. Many shade-demanding plants have been cultivated by horticulturists for use in shaded areas or indoor planters.

GROWTH INHIBITORS are produced in the leaves of some plants and are released into the soil or air. They become distributed in a widening circle around the base of the plants. These substances stop or retard the competitive growth of their own seedlings as well as those of other species. In this way the parent plant can utilize the available moisture and nutrients with the least competition. Man has taken advantage of growth inhibitors for bacteria and fungi by adding them to his arsenal of antibiotic drugs.

FUNGI are involved in unique relationships with a number of kinds of plants. Fungi that grow in and around the roots of some kinds of forest trees and orchids effectively improve the capacity of the roots for absorbing moisture and nutrients. In turn, the fungi get some of their nutrients from the host plant's roots. The host plants cannot survive without the fungi.

Lichens are a partnership of fungi and algae. Pooling resources enables these plants to occupy bare rocks, polar regions, and other hostile niches.

In shade of tree, lawn grasses grow poorly or not at all.

ANIMAL-TO-ANIMAL INTERACTIONS are equally diverse, further illustrating the delicate biological balances that have developed in ecosystems. If these relationships are disturbed, the effects may be magnified remote from the disruption.

PREDATOR AND PREY and parasite and host are obvious interactions between animals. In these associations, one animal benefits, the other is harmed.

Mimics that resemble a poisonous or unpalatable species may confuse a predator and thus be spared. Some kinds of beetles not only resemble ants but also behave like them. With this protective resemblance, they live and feed in an ant colony without being rejected or preyed on themselves.

COMPETITION between similar species or populations may be so intense that resources of the habitat are not adequate to support both. Sometimes one of the species may adjust to the pressure by changing its food habits. Or one species may crowd out its competitor by sheer numbers. It is an axiom that two similar species cannot coexist permanently in a niche.

IN SYMBIOTIC RELATIONSHIPS, both species benefit. Animals that graze on tough prairie plants have special populations of microorganisms living in their digestive tracts. These assist in breaking down foods for assimilation, and can thrive only in the rich food surroundings of the host's digestive tract.

Man is no exception to such a relationship. Antibiotics taken to destroy infections may also kill the microorganisms in his intestines. Then he is unable to digest food properly until a new population becomes established. Termites depend on the protozoans in their intestinal tracts to digest the wood they eat.

Cowbirds of temperate regions and some of the weaverbirds of the tropics lay eggs in the nests of other birds. The host female incubates the egg (or eggs) and cares for the young. Her own young may or may not be starved or crowded out of the nest by the larger and generally more aggressive "parasite." This relationship is more imposition than parasitism or symbiosis.

An otter catching a fish: a predator-prey relationship.

PLANT AND ANIMAL INTERACTIONS represent many variations and levels of relationships. Commonly they involve food for the animal and distribution of seed or pollination for the plant. Some of these relationships change with the seasons or with different life stages.

A GIVE-AND-TAKE RELATIONSHIP exists between the trees in a forest and the squirrels and other rodents that feed on their seeds and nuts. Because they eat some of the annual crop, the animals tend to limit the production of seedlings. This is counterbalanced at least partly by their inadvertent "planting" of some seeds and nuts that the animals put in storage or hiding for their own use. Those buried in favorable places for germination and growth produce seedlings.

Seeds or nuts carried far from the parent plant increase the range of the plant species. They also reduce the competition that would occur if they grew near the parent plant. Later the animals may use the matured plants for dens or nests.

DEPENDENCY of plants on animals to pollinate their flowers is a direct interaction. Bees, wasps, flies, moths, butterflies, and some birds and bats are the chief animal pollinators. Some flowers are extremely modified to accommodate or to attract their pollinators, for whom both the nectar and the pollen may serve as food. Even the opening of the flowers may be synchronized with the flight habits of the pollinator. Without these animals to pollinate them, some plants could not bear seeds and would disappear in time. Likewise, some pollinators depend on the plants for food or living space.

PROTECTIVE relationships have evolved between ants and some species of plants. An ant that lives in the hollow base of thorns of an acacia in South America feeds on the plant's sweet secretions. The ants swarm out of the thorns and drive away intruders by stinging them.

Leaf-cutting ants bite out small pieces of leaves and carry them to their nests. In extensive underground gardens, they mix the chewed leaves with the filaments of fungi that they grow there. The fungi absorb the leaf material and grow profusely. The ants in turn eat the fungi.

Leaf-cutting ants interact with plants, growing fungi on leaves.

BASIC MATERIALS

Both the living and the nonliving parts of the environment consist of basic materials called elements. Of the 105 elements so far identified, only 15 to 20 are consistently involved in supporting life on earth. These are called the essential elements, and they occur in all four of the interrelated spheres of the earth: the atmosphere, or air; the hydrosphere, or water; the lithosphere, earth's crust; and the biosphere, where life exists on earth (p. 5).

Earth evolved slowly to its present capacity for supporting large amounts of life, for increases in life systems were possible only as the essential elements became available from the abiotic system. In the beginning, several of these elements were in short supply or were not usable by living things in the form in which they occurred. As life evolved, cycles also developed that kept these essential elements available on a re-use

PERIODIC TABLE OF ELEMENTS

H Hydrogen								
Li Lithium	**Be** Beryllium							
Na Sodium	**Mg** Magnesium							
K Potassium	**Ca** Calcium	**Sc** Scandium	**Ti** Titanium	**V** Vanadium	**Cr** Chromium	**Mn** Manganese	**Fe** Iron	**Co** Cobalt
Rb Rubidium	**Sr** Strontium	**Y** Yttrium	**Zr** Zirconium	**Nb** Niobium	**Mo** Molybdenum	**Tc** Technetium	**Ru** Ruthenium	**Rh** Rhodium
Cs Cesium	**Ba** Barium	**La** Lanthanum	**Hf** Hafnium	**Ta** Tantalum	**Wa** Tungsten	**Re** Rhenium	**Os** Osmium	**Ir** Iridium

The essential elements—needed by nearly all plants and animals.

These elements are needed in lesser amounts by both plants and animals.

Elements needed by some but not all organisms.

plan. The elements absorbed by living things are not locked permanently into the biotic or life systems but are returned again and again to the abiotic system.

Ecologists make intensive studies of these biogeochemical cycles, for they are the mechanism by which life and earth are bound together. They must have a keen understanding of how these cycles work, for life depends on a continuous supply of essential elements. As one example, life cannot exist without water, which is one important source of two of the essential elements —hydrogen and oxygen. All living cells contain proteins that consist of hydrogen, oxygen, carbon, nitrogen, sulfur, and phosphorous. The chlorophyll of green plants, on which all life depends, is composed of hydrogen, oxygen, carbon, nitrogen, and magnesium. Other such examples could be given to prove unquestionably the necessity of essential elements. Some of the important cycles are described on the following pages.

There are 105 known elements. Not all are shown on this chart.

								He Helium
	symbol name		B Boron	C Carbon	N Nitrogen	O Oxygen	F Fluorine	Ne Neon
			Al Aluminum	Si Silicon	P Phosphorus	S Sulfur	Cl Chlorine	Ar Argon
Ni Nickel	Cu Copper	Zn Zinc	Ga Gallium	Ge Germanium	As Arsenic	Se Selenium	Br Bromine	Kr Krypton
Pd Palladium	Ag Silver	Cd Cadmium	In Indium	Sn Tin	Sb Antimony	Te Tellurium	I Iodine	Xe Xenon
Pt Platinum	Au Gold	Hg Mercury	Tl Thallium	Pb Lead	Bi Bismuth	Po Polonium	At Astatine	Rn Radon

MICRONUTRIENT ELEMENTS are those needed in organisms in only trace amounts, but they are as essential as the macronutrient elements. The law of the limiting factor (p. 143) applies.

Boron, chlorine, cobalt, copper, iron, manganese, molybdenum, and zinc are among the micronutrients. In ecosystems, they are kept in cycle by natural decay processes and then reabsorption by plant roots.

Some micronutrients are involved with enzymes in physiological processes. Deficiency symptoms and malfunctioning occur if these elements are not available in sufficient quantity or in a usable form. An entire ecosystem can fail if organisms near the base of the food chains are affected by this lack. Human health problems have been traced to a micronutrient deficiency in the soil on which crops are grown. If micronutrients occur in greater than trace amounts, they are usually extremely toxic. For this reason, they must also be monitored where pollution has become a problem.

Iron is needed by both plants and animals. Leaves cannot turn green and produce food unless supplied with trace amounts of iron. Without iron, hemoglobin cannot form in the blood. ▼

Iodine is a constituent of a hormone of the thyroid glands. Without iodine, thyroxin cannot be produced; the thyroid enlarges, producing a goiter. Use of iodized salt prevents this. ▼

plus iron   minus iron

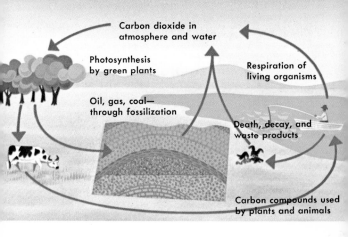

Carbon dioxide in atmosphere and water

Photosynthesis by green plants

Respiration of living organisms

Oil, gas, coal— through fossilization

Death, decay, and waste products

Carbon compounds used by plants and animals

CARBON occurs in all organic compounds, hence the carbon cycle is in a key position in the biosphere. It is interlocked with several other cycles.

Combined with other elements, carbon is used and stored in many ways. All living things are temporary storage places for carbon. Fossil fuels consist of carbon compounds that were produced by photosynthesis millions of years ago. Carbon is stored in large amounts in the wood of trees by the same process today.

ATMOSPHERIC CARBON is largely carbon dioxide (CO_2), a gas that represents about 0.03 percent of the atmosphere by volume. This carbon is the important energy link between the sun and the earth, for by changing chemically, it acts as the recipient of the sun's energy for living systems. Enriched with energy from the sun, it is built into food compounds in the photosynthetic process in green plants. The energy contained in these compounds can be released from them at low temperatures by respiration or at high temperatures by rapid oxidation or burning. By these processes, carbon dioxide is again released to atmosphere.

Photosynthesis and respiration are the route by which carbon atoms are reused to sustain life. It has been calculated that in this manner atmospheric carbon is recycled in the biosphere about every 300 years.

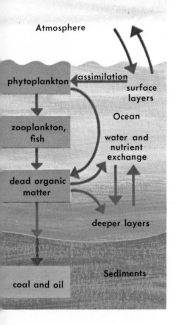

Atmosphere

phytoplankton — assimilation

surface layers

Ocean

zooplankton, fish

water and nutrient exchange

dead organic matter

deeper layers

coal and oil

Sediments

HYDROSPHERE CARBON is an enormous reservoir. Photosynthesis by aquatic plants and respiration by both plants and animals involves the same chemistry as for land plants and animals, hence carbon is abundantly available in water either as a gas (CO_2), as carbonates, or in organic compounds.

A constant adjustment takes place between the carbon dioxide dissolved in the water and that which is in the atmosphere. Biologists estimate that about half of the carbon dioxide produced and released into the atmosphere is absorbed by the oceans. As the marine organisms die, some of them sink into deep water where the lack of oxygen prevents decay. Organic carbon thus collects in large amounts in ocean bottom sediments.

A series of chemical reactions, the carbonate-bicarbonate system, also occurs between carbon dioxide and water. More carbon is contained in this system than in the atmosphere. Limestone ($CaCO_3$) is precipitated from this system, hence a great amount of carbon is locked in the world's limestone deposits. It is released slowly when the limestone is exposed to other chemicals that are derived from life processes.

PHOTOSYNTHESIS is dependent on carbon dioxide. On warm, sunny days, the carbon dioxide level in the air near the ground commonly decreases by late morning. This limits the rate at which photosynthesis can occur, thus reducing for the remainder of the day the amount of basic food being produced. The normal recharge of carbon dioxide comes at night when the gas is being released by both plants and animals in the respiration process and no photosynthesis is occurring.

Some scientists have suggested that the rate at which carbon dioxide has been released into the atmosphere from the burning of fuels in recent decades might cancel this natural limitation on the rate of photosynthesis. This effect has been observed only experimentally, not in nature.

SEASONS affect the carbon dioxide content of the atmosphere. In northern latitudes, there is normally an increase in carbon dioxide during the winter. At this time of the year, photosynthesis is restricted largely to the evergreen trees. In areas that are dominated by deciduous trees, leafless in winter, the release of carbon dioxide by respiration is greater than the amount utilized by photosynthesis. This effect is short-lived because of the fairly rapid mixing of atmospheric gases and also the great capacity of water to absorb carbon dioxide from the atmosphere.

GREENHOUSE EFFECT refers to the warming of the earth's surface due to the accumulation of carbon dioxide in the upper atmosphere. The carbon dioxide molecules absorb and trap infrared radiation (heat) that would otherwise escape the atmosphere. Carbon dioxide may be increasing in the upper atmosphere due to the burning of fossil fuels, a subtle but consistent effect that man may have on the earth's climate. A few degrees change in the yearly average temperature could melt the polar icecaps and cause extensive changes in sea levels.

Layer of carbon dioxide traps infrared radiation and returns it to surface again. This is called the greenhouse effect.

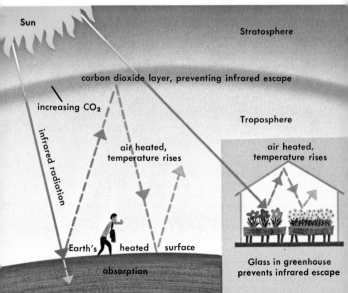

Sun

Stratosphere

carbon dioxide layer, preventing infrared escape

increasing CO$_2$

Troposphere

infrared radiation

air heated, temperature rises

air heated, temperature rises

Earth's heated surface

absorption

Glass in greenhouse prevents infrared escape

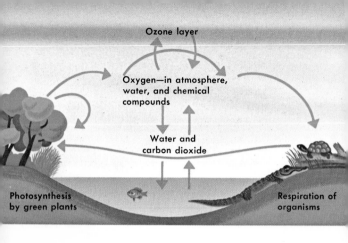

Ozone layer

Oxygen—in atmosphere, water, and chemical compounds

Water and carbon dioxide

Photosynthesis by green plants

Respiration of organisms

THE OXYGEN CYCLE is extremely complex and is interwoven with several other biogeochemical cycles, particularly those involving carbon, hydrogen, and water. Life is impossible without a continuous supply of oxygen, the ultimate source of which is photosynthesis.

Molecular or free oxygen (O_2) is a gas used by nearly all organisms in respiration. Its importance in the biosphere is very direct. Water, carbon dioxide, oxides of other elements (particularly metals), and most all organic material are among the sources of oxygen. In addition to molecular oxygen, the atmosphere also contains ozone (O_3) and atomic oxygen (O).

ATMOSPHERIC OXYGEN occurs in the lower levels of the atmosphere, where most life exists, at about 20 percent by volume. At one time, ecologists were concerned that man's increased use of fossil fuels would reduce the amount of atmospheric oxygen to a critical level for the existence of living things. Recent calculations indicate that this is not a great danger. Surveys started in the early 1900's show that the percentage of oxygen in the atmosphere has not varied significantly over the years. Studies also indicate that oxygen atoms are recycled through the biosphere by photosynthesis about every 2,000 years.

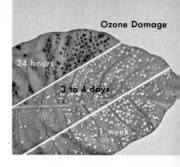
Ozone Damage

24 hours

3 to 4 days

1 week

HYDROSPHERIC OXYGEN is most important for aquatic organisms. Water normally contains a large reserve of oxygen. The amount that can be dissolved in water is determined by temperature and by pressure. Oxygen passes from the atmosphere to the hydrosphere, or vice versa, until an equilibrium is established. At saturation, the water contains all the oxygen it can hold.

If photosynthetic conditions are optimum for aquatic plants, water may become supersaturated with oxygen. The excess is then released into the atmosphere. "Kills" of fish and other aquatic organisms commonly occur because of oxygen depletion, a condition that is most likely to develop in warm waters when the capacity of water to hold oxygen is low and on cloudy days when the rate of photosynthesis is slowed.

LITHOSPHERIC OXYGEN occurs between soil particles in air spaces of varying sizes. For most roots and organisms to exist in soil, it is essential that oxygen be present in these spaces. Under prolonged flooding, the spaces fill with water, causing a drop in the amount of available oxygen. Most forms of life, even large trees, die without this oxygen supply.

The lithosphere also contains huge quantities of oxygen as oxides and carbonates in ores, rocks, and mineral soil particles. It can be released only by chemical changes.

OZONE (O_3) is both beneficial and harmful to life. A natural region of ozone in the upper atmosphere absorbs part of the ultraviolet radiation from the sun, forming a protective shield that prevents harmful radiation from reaching plants and animals. The basic source of the ozone is the molecular oxygen (O_2) produced by photosynthesis.

In recent decades, the amount of ozone in the lower atmosphere has been increasing due to photochemical action of sunlight on auto exhaust chemicals. Even small amounts of ozone can cause serious damage to plants. This ozone is part of the smog complex that affects large population centers.

relatively constant

| O_2 20.9% | IN ATMOSPHERE | |
| CO_2 0.03% | N 78% | |

varies

| O_2 0-21% | IN SOIL | |
| CO_2 0.03-21% | N 78% | |

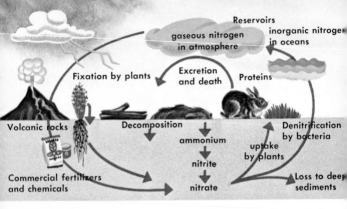

Reservoirs
gaseous nitrogen in atmosphere
inorganic nitrogen in oceans
Fixation by plants
Excretion and death
Proteins
Volcanic rocks
Decomposition
Denitrification by bacteria
ammonium
uptake by plants
nitrite
Commercial fertilizers and chemicals
nitrate
Loss to deep sediments

NITROGEN, one of the essential elements, is part of every molecule of chlorophyll and protein, hence is necessary to the food-making process. As a free or inert gas, nitrogen comprises about 79 percent of the atmosphere, but in this gaseous state, it is not usable by most forms of life. For life processes, the nitrogen must be "fixed"—occurring in such compounds as ammonia, nitrates, urea, amino acids, or proteins. The abundance of life at a particular time can be limited by the amount of these usable nitrogen compounds in the ecosystem.

HUMAN POPULATIONS that are successful—that is, prosperous and well fed—depend on a highly productive agriculture which, in turn, is based on man's ability to make artificial fertilizer, thus fixing nitrogen chemically. This fertilizer has made possible great yields of food even from poor land. Natural nitrogen fixation is not adequate for producing the enormous amounts of food needed to sustain the world's rapidly increasing human population.

Man's food production and his sewage are now upsetting the natural nitrogen cycle. Fertilizers reach waterways and cause excessive aquatic plant growth. Incompletely treated sewage is also a source of an overabundance of these nutrient compounds and elements. These additions overpower the ecosystem's capacity to remain regulated. Fish kills occur, and the water becomes unfit for most living things. These are indications of eutrophication (p. 118).

MICROORGANISMS of several kinds are involved in the natural processes of nitrogen fixation. Others release the fixed nitrogen into the atmosphere as free gaseous nitrogen, which is denitrification. A few abiotic processes, such as lightning, add small amounts of fixed nitrogen to the ecosystem. Thus there is a continuous recycling of nitrogen between the lithosphere, biosphere, hydrosphere, and atmosphere. This cycle is closely coordinated with such other ecosystem cycles as those of carbon and oxygen.

In symbiotic nitrogen fixation, the microorganisms that can fix nitrogen live within the tissues of the free-living host plants that are rooted in the soil. Bacteria, phycomycetes (algal fungi), and blue-green algae are among the several kinds of lower plants involved in the process. The best known higher plant hosts are alfalfa, clovers, peanuts, and other legumes. Several other kinds of plants also act as hosts. Among lesser known hosts are wax myrtles, Ginkgo, alders, New Jersey tea, cycads, and podocarps.

In most cases, the symbiotic microorganisms invade the host's root system, but they are found also in the leaves and stems of some plants. When the host dies, the fixed nitrogen is released to the soil during decay. Non-symbiotic plants can then absorb this fixed nitrogen.

The amount of nitrogen fixed biologically is significantly less than what is needed to produce high yields of crops. The earth's nitrogen cycle was naturally balanced before the intrusion of man, and so ecosystems were tuned to the smaller amounts.

Blue-green algae on soil

Nodules in wild coffee leaf

Root nodules on legume

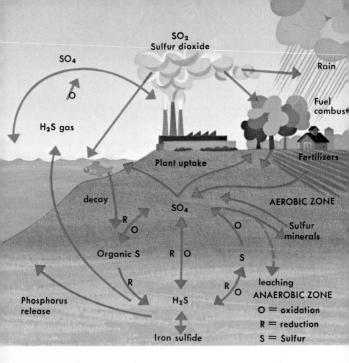

SO₂
Sulfur dioxide

SO₄

Rain

H₂S gas

Fuel combust*

Plant uptake

Fertilizers

decay

SO₄

AEROBIC ZONE

Organic S

Sulfur minerals

Phosphorus release

H₂S

leaching
ANAEROBIC ZONE
O = oxidation
R = reduction
S = Sulfur

Iron sulfide

SULFUR enters food chains by being absorbed as a sulfate by plants. There it is incorporated into amino acids, which are the basic building blocks of the proteins in all living things.

Most of the sulfur in the biosphere is not in a form that can be utilized by living things. These forms—pure sulfur, sulfur dioxide, hydrogen sulfide, metallic sulfides, and other compounds—represent the reserve pool of sulfur. These forms of sulfur are toxic to many kinds of life, but some kinds of bacteria can utilize these forms in their sulfur cycles.

THE SULFUR CYCLE links the available and the unavailable forms of sulfur and allows a continuous replenishment of the available sulfate form. Microorganisms are responsible for the recycling of sulfur between the reservoir and the biosphere.

Proteins in the excretions of organisms and in their dead bodies are converted into sulfates by the bacteria and fungi. These sulfates can be returned to the food chain immediately by being absorbed and processed by green plants. Oxygen must be present for this conversion, however. In deep water and in muds, where oxygen is lacking, other kinds of bacteria reduce these materials to pure sulfur and to sulfides.

HYDROGEN SULFIDE is commonly produced in anaerobic (lacking oxygen) bottom muds. Its odor, like rotten eggs, can be smelled when these waters and muds are stirred. When such large amounts of sulfates are released that the biological systems become overloaded, so much poisonous hydrogen sulfide may be produced that massive fish kills occur.

Safeguards against the accumulation of hydrogen sulfide are built into the sulfur cycle, for colorless sulfur bacteria and some of the green, photosynthetic bacteria can use both hydrogen sulfide and pure sulfur in their metabolism. Harmful sulfides may also be precipitated with iron as iron sulfide when the two elements occur together in anaerobic muds.

Sulfur dioxide (SO_2) damage to apple leaves.

SULFUR DIOXIDE is one of the principal air pollutants, causing damage to plants as well as being a component of smogs (p. 116). It now occurs in the atmosphere in such large quantities that the sulfur cycle may be overloaded, but the effects are not yet fully known. The final absorbers of most of this harmful gas are apparently the soil and the oceans, where it is taken into cycle again.

Discoloration of paint on boat is due to hydrogen sulfide in water.

PHOSPHOROUS now in cycle in ecosystems originates from deposits formed in rocks in the geologic past. This phosphorous is added to natural systems very slowly. As a result, the phosphorous needed by living things comes primarily from a recycling of the limited amount that is already found in living and nonliving organic materials.

Some ecologists contend that a phosphorous deficiency is the most important factor in limiting the productivity in many ecosystems. With hydrogen, oxygen, nitrogen, and sulfur, phosphorous is an essential component of proteins, the main ingredient of protoplasm. Phosphorous is also a part of DNA, the molecule that directs cell activities, and of ATP, the energy-transferring molecule. No life can exist without an available supply of phosphorous for these important needs.

Phosphorous is lost constantly from food chains as organic matter becomes a part of the sediments on the bottoms of deep lakes and in the oceans. Only major upwellings can bring this phosphorous back into circulation again. This does occur regularly in some lakes and in specific areas of the seas. Rich "blooms" of algae and other microscopic forms of life generally follow the occurrence, and good fishing then follows. To support his crops and livestock, man produces large amounts of phosphate fertilizers (p. 34).

PHOSPHOROUS POLLUTION is a relatively new problem. Detergents containing phosphates are released in great quantities into natural waters as wastes. In these environments, phosphorus is no longer a limiting factor, and as a result, a population explosion of plants, or "bloom," occurs upsetting the natural ecosystem.

Methods of removing phosphates and nitrogen from wastes before their discharge are being improved constantly. One promising technique is to pump partially treated sewage water onto poor land. This fertilizes as well as waters the soil and generally results in better and increased crop production.

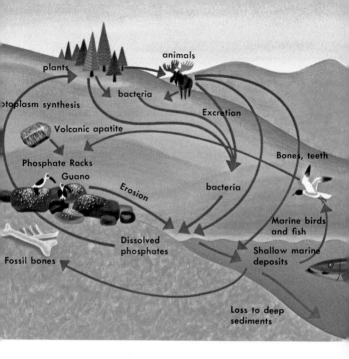

CYCLING OF PHOSPHOROUS in aquatic ecosystems has been studied extensively. The total phosphorous present is the sum of three forms: inorganic, particulate organic, and dissolved organic. The first, known chemically as orthophosphate, is the form that most plants absorb, use, and then change into the second form. Consumers get their organic phosphorous from their plant or animal food. Dissolved organic phosphorous is produced by the excretions of living organisms and by the decay of their bodies.

TURNOVER TIME refers to the length of time that elapses for phosphorous (and other elements) to be absorbed, utilized, and then returned to its absorbable inorganic state. Obviously, if this time is short enough, a relatively small amount of phosphorous can sustain an ecosystem. Studies made with radioactive phosphorous indicate that the rate is indeed rapid, ranging from a matter of minutes up to twelve days or longer. The rate differs with the ecosystem and also with the season, being longest in winter.

SOILS, a source of other nutrients, are usually considered to be abiotic or nonliving, but they are really almost living. A barren area, for example, is said to have "dead soil."

Soils are complex, changing natural bodies that possess both biotic and abiotic characteristics. Minerals, water, and air constitute the abiotic portion, while the biotic components are the organisms that live on and in the soil and the organic matter that they produce and then recycle. To understand an ecosystem, it is necessary to know what soils are, how they develop, how they relate to food chains, and what lives in them.

MATURE SOILS have distinctive profiles. The top horizon is "A," the middle "B," and the third is "C"—closest to parent material.

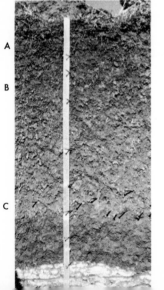

PLANT ROOTS get from the soil most of the essential elements and water that enters the food webs at the producer level. Though often ignored, the extensive network of plant roots in the soil is an important part of the total soil biomass.

Soils are also the habitat for numerous kinds of invertebrates and microorganisms, plus a few kinds of higher animals. Most soil organisms are too small to be seen without a microscope, and so their numbers and biomass are usually surprising. In fertile soil, a teaspoon of soil may contain from 5 to 6 billion bacteria, protozoa, nematodes, and other small creatures. It is estimated that the population of these organisms in a square foot of forest floor may be 40 times greater than the total human population in the world.

Nearly all are very active in the detritus food chain. They play a significant role in the recycling of minerals and in decay processes.

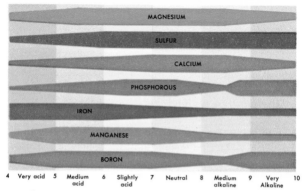

| 4 Very acid | 5 Medium acid | 6 Slightly acid | 7 Neutral | 8 Medium alkaline | 9 Very Alkaline | 10 |

Availability of Nutrients Related to pH 4 to 10

SOIL CHEMISTRY is also important, for it is the combination of the chemistry of the soil and biological activity that makes possible the cycling of minerals.

A soil's neutrality, acidity, or alkalinity is represented by the symbol pH. The most productive soils are neither very acid nor very alkaline. Sometimes when the pH shifts, the essential elements may still exist in the soil but not in a form that can be absorbed by plant roots. For this reason the elements cannot be moved out of the soil and into the food chains.

The chart above shows the relationship of pH to the availability of some of the essential elements. Availability is indicated by the width of the band of color that represents the element. In agriculture, the liming of soils drives the pH toward the alkaline side, while adding organic matter tends to move it toward the acid side. Farmers use this system to obtain the maximum benefit from their fertilizers. In natural ecosystems, the pH of the same kind of soil may change from time to time and from place to place.

Habitats are the places where plants, animals, and microorganisms live. Within a habitat, each organism fits into a special niche that is determined by how it lives or what it does.

Largemouth bass and bullheads, for example, are aquatic animals. Both can inhabit the same warm-water lake or pond. But the bass, a predator, feeds mostly on smaller live fish and spends most of its time in the open water while the bullhead lives close to the bottom where it feeds mostly as a scavenger. The two fish occupy different niches in the same habitat.

Similarly, slash pine and bald cypress can grow next to each other. But the slash pine grows best on relatively higher, drier land, while the bald cypress grows on nearby, moist or swampy soil. Each has its niche within the same habitat.

The broadest categories of habitats are terrestrial (land) and aquatic (both fresh and salt water). These are the major macrohabitats. Two lesser habitats, subterranean and aerial, are also treated here.

SUBTERRANEAN HABITATS include caves and underground cavities, with the water they contain. Soil or loose surface debris in which animals burrow and roots grow are also subterranean habitats. Some animals inhabit these places only occasionally—hiding, resting, hibernating, or rearing young there. Among these are swifts, swallows, and South American oilbirds. Bats are the best known of the part time cave dwellers, hundreds of thousands of them congregating in some large caves. Their guano is a major energy input in food chains of permanent cave inhabitants.

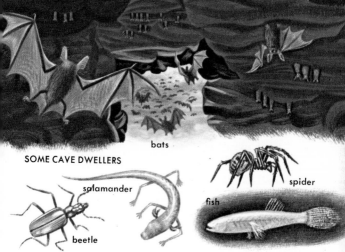

bats

SOME CAVE DWELLERS

beetle

salamander

spider

fish

CAVE-DWELLING ORGANISMS include some species of salamanders, fish, snails, and worms, but the most numerous are the arthropods—the insects, spiders, and crustaceans. Bacteria and fungi, which do not require light, are the only plants that grow in caves. They subsist on the dead remains of animals or on plant debris from the surface.

All except a few caves are very moist, and the dry ones do not have many inhabitants. In this high humidity the insects, snails, and other animals have thinner outer coverings than do surface relatives, for there is less danger of drying out.

Caves are uniformly cool the year around, and there is little or no daily temperature fluctuation. With literally no "seasons," cave animals breed throughout the year. In their isolation from surface predators and competition, many types of cave animals have persisted as living relics of groups best known as fossils.

True cave animals react negatively to light. With few exceptions, they are white, colorless, or at least much paler than their nearest relatives that live on the surface. Some kinds will regain their pigments in only a few months if they are brought to the surface. Others die if exposed to light for more than a brief interval. Cave dwellers have small or no eyes, but antennae, tactile hairs, and other sense organs are usually well developed.

Simplified Food Chain

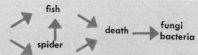

guano → insect → fish → death → fungi bacteria

spider

AERIAL HABITAT implies that the plants or animals live in the air permanently. Some kinds of microorganisms do inhabit the air, but for most living things, the air is only a temporary habitat.

BATS are the only mammals that have wings and are capable of true flight. Most bats are nocturnal insect eaters, collecting their food as they fly. During the day, they roost in dark places.

Some birds are "on the wing" most of their lives, settling to earth only to nest and rear young. Among these masters of the air are albatrosses, petrels, sooty terns, and other seabirds.

INSECTS rank next to birds in flying skill. In geological history, they were the first animals capable of flight. Most flying insects are rather fragile, remaining in the air for only short distances and duration. But some do travel astonishingly far. Painted lady butterflies migrate from California to Hawaii— more than 2,000 miles.

OTHER KINDS OF ANIMALS can glide, stretched membranes of skin serving as "wings." Among these are flying squirrel and several Asiatic species of snakes, lizards, and frogs.

Young spiders "balloon" into new territories. They let out one or several strands of silk that are caught by the wind and pull the spider into the air. Often they may ride for many miles.

"AIR" PLANTS, or Epiphytes, are plants that occupy a type of aerial niche. Among these are some kinds of orchids, cacti, bromeliads, ferns, mosses, liverworts, and algae. They grow on tree trunks, roofs, wires, or other supports without need for contact with the soil. Their inorganic nutritional needs are supplied largely from the air.

SOME TEMPORARY AIR INHABITANTS

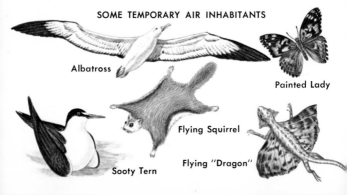

Albatross

Painted Lady

Flying Squirrel

Sooty Tern

Flying "Dragon"

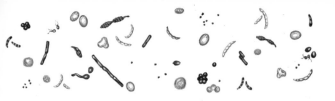

This simple air sampler traps airborne microorganisms in the liquids in the bottom of flasks.

This field trap collects spores on microscope slides to detect diseases (blights, mildews, etc.) for early control measures.

MICROORGANISMS exist in the air near the earth's surface in a surprising abundance. Even on a clear day, the air contains enough solid materials to provide niches for the existence of microorganisms. Algae, protozoans, and bacteria are commonly collected in air samples. They must be screened from hospital operating rooms and manufacturing processes where absolute freedom from aerial contamination is necessary.

The microscopic parts of plants and animals also play significant ecological roles in the air. Spores of ferns, mosses, and fungi are distributed through the air regularly, as are the pollen grains of many seed-bearing plants. They are essential for the reproduction and survival of these species in their ecosystems.

Spores and pollen in the air are major factors in the human allergy called hay fever. Bits of insect wings and other animal parts may also be airborne. The stinging scales of moths drifting in the air are a plaguing nuisance in some parts of the tropics. Although each plant or animal part or microorganism is small, they add up to tons of material. Their circulation through the air contributes significantly to the relocation and recycling of organic material on the earth's surface.

TERRESTRIAL (LAND) HABITAT supports about 80 percent of all the species of plants and animals that inhabit the earth. In this habitat they have achieved their greatest variety and most advanced development. The land habitat provides animals with the greatest abundance of plant food with a minimum of competition.

OXYGEN is the great luxury of the land habitat. On an average, it is available in a quantity about 200 times greater than in water. This permits land animals to function at a peak of metabolism. This abundance of oxygen is true only in the thin layer close to land, however. Animals that venture into the air are limited by the lack of oxygen at high altitudes.

LOW HUMIDITY is the greatest disadvantage in the land habitat. This has brought about the development of special coverings to prevent drying out. Land plants,

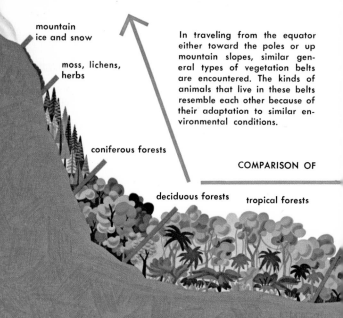

mountain ice and snow

moss, lichens, herbs

coniferous forests

deciduous forests

tropical forests

In traveling from the equator either toward the poles or up mountain slopes, similar general types of vegetation belts are encountered. The kinds of animals that live in these belts resemble each other because of their adaptation to similar environmental conditions.

COMPARISON OF

for example, are covered with a nearly water-impervious layer called the cuticle. The most successful of the land inhabitants have also developed firm skeletons or supporting structures, for compared to water, air has a much lower density and does not buoy their bodies.

Living conditions on land range from basins below sea level to mountain peaks that are more than five miles above sea level. Between these two extremes and at all altitudes are hills, valleys, and plains. The soil of the earth varies from sands and rocks to muds and clays. Some soils are rich with humus, the decayed remains of plants and animals.

THE TEMPERATURE in polar regions seldom gets above freezing at any time of the year, while in deserts the air temperature may rise to over 130 degrees F. and even higher in pocketed areas. Both at the poles and in the tropics, the temperature is fairly constant throughout the year; in temperate regions, it fluctuates widely seasonally and also daily. This, too, is a variation from most aquatic environments in which the temperature is either uniform the year around or changes very slowly.

RAINFALL is another factor of the land habitat. Tropical rain forests have an annual rainfall of from about 90 inches to several hundred inches, while in hot deserts the rainfall may be less than an inch a year.

Despite the extremes of the land habitat, similarities in conditions at the same latitudes and altitudes result in striking ecological duplications. Plants and animals in these locations may develop almost identical forms and structures even though they may be unrelated. These are most obvious and most clearly observed in the broad associations of plants and animals called biomes (p. 54).

VERTICAL AND HORIZONTAL LIFE ZONES

deciduous forests

coniferous forests

herbs, lichens, mosses

polar ice and snow

SALT WATER HABITATS cover 70 percent of the earth's surface, the most extensive of the major habitats. More different phyla of animals live in the sea than on land. Within limited areas, the seas are much richer with life than are equal areas on land. Yet there are fewer species in the seas, marine organisms accounting for only 20 percent of all species. Because of the generally uniform conditions and the lack of isolation in the continuous water medium, seas do not favor the development of species as much as does land.

Only a few kinds of flowering plants live in the sea. Marine plants are mainly algae, and the most numerous of these are microscopic species that form the vast floating pastures of phytoplankton. These small plants account for most of the photosynthesis in the sea and form the base of the food chain for marine life. Seaweeds that grow in shallow seas are macroscopic types of algae with fleshy bodies. Some of them, the giant kelps, grow to more than 200 feet long. They are anchored to the bottom by rootlike "holdfasts," and their fleshy, leaflike bodies are floated at the surface by air-filled bladders. Where abundant, these giant seaweeds make "forests" in the sea.

Despite its immensity, the sea is a relatively uniform habitat. Though there are local variations, the salt content is fairly constant—about 3.5 percent. Tropical seas have an average temperature of about 85 degrees F., while in arctic seas the temperature is always near freezing. But in the same area, the temperature of the water in the sea rarely varies much more than 5 degrees from season to season.

The oceans are divided into several distinct kinds of habitats. The major divisions are described here and on the following pages.

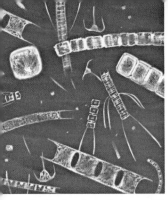

Planktonic algae (microscopic) are the basic food in open sea.

holdfast for kelp

Kelps are giant algae that grow in shallow areas.

THE INTERTIDAL ZONE is the area close to shore that is alternately exposed to the air or is covered by water as the tides move in and out twice daily. This is the beach region, which may be sandy, rocky, or muddy. On some shores, the waves roll in smoothly and rhythmically; on others, they pound.

This zone is rich with both food and oxygen, but it is never-theless extremely difficult for plants and animals to live in because of the constant movement of the water. Some kinds of animals burrow into the mud or sand when the tide leaves. They remain there until the water floods the shore. Other animals and plants cling tenaciously as strong waves sweep over them. Special structures are required for holding and for protection.

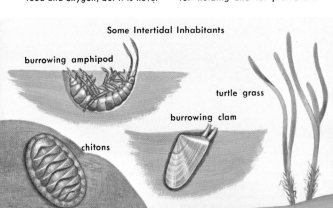

Some Intertidal Inhabitants

burrowing amphipod

turtle grass

burrowing clam

chitons

THE NERITIC ZONE is the area of the sea over the continental shelves. Both the intertidal zone (p. 47) and coral reefs (p. 64) are subdivisions of the neritic zone. The maximum depth of the water is about 600 feet. This is the absolute limit for light penetration in amounts sufficient for photosynthesis to occur and only reaches such depths in clear, tropical waters.

Wave action is detectable throughout the neritic zone, and life occurs in abundance and great variety from the surface to the bottom. The peak of productivity is in areas of upwellings, where vertical currents lift nutrition-rich waters from the bottom to the surface. These waters over the continental shelves represent less than 10 percent of the ocean surface, but they are the main source of commercial fish harvests.

In areas enriched by upwellings, "blooms" of plankton may occur. Generally they provide more food. Florida's destructive red tides, however, result from the bloom of a dinoflagellate, *Gymnodinium*, that gives off a poison killing fish by the hundreds of thousands.

The *bathyal zone*, from 600 to 6,000 feet on the continental slopes, represents the beginning of the deep sea. Here there is only a slight movement of the water and only traces of light in the upper level.

The *abyssal zone*, below 6,000 feet, is an area where the water is uniformly cold, quiet, and perpetually dark. Most of the oceans are, in fact, very deep—averaging more than two miles. It is astonishing that living things—mostly animals—do exist in the greatest depths, more than seven miles below the surface.

Deep-sea animals are typically small but have large mouths and a pouchlike stomach that enables them to gorge when food is available. Most of them are predatory, feeding on animals that they capture in the darkness. Others are scavengers, making meals of the dead plants and animals that sink from the lighted zones above.

Taken from Gemini spacecraft, this photo of tip of Taiwan shows upwellings (dark blue) along upper east and west coasts. In these food-rich areas, production of fish is greatest, hence the fishing is good.

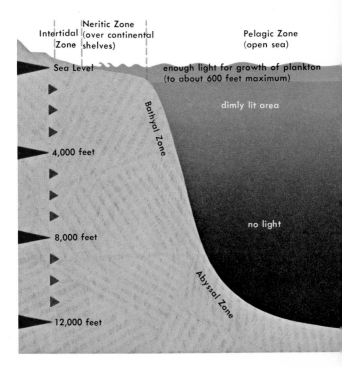

Intertidal Zone | Neritic Zone (over continental shelves) | Pelagic Zone (open sea)

Sea Level

Bathyal Zone

enough light for growth of plankton (to about 600 feet maximum)

dimly lit area

4,000 feet

8,000 feet

no light

Abyssal Zone

12,000 feet

THE PELAGIC ZONE is the open sea. In its upper region—above 600 feet—are vast quantities of plankton, plants and animals that form the base of the food chain in the sea. Many kinds of fish are equipped to strain these small morsels of food from the water through sievelike meshes on their gills.

Most animals that live in the open sea are streamlined for swift swimming or are adapted for floating as jellyfish do. This is in contrast to the many kinds of sessile or attached types that live in shallow waters and depend on the currents to bring them the sea water containing their food.

The largest populations of a single species, such as the herrings, occur in cold seas. Tropical and subtropical seas contain a more varied population but fewer individuals of any species.

49

FRESHWATER HABITATS may be either the running waters of streams (lotic) or the standing waters of ponds, lakes, swamps, and marshes (lenitic). Compared to the seas, this is a very small habitat. Totally, fresh waters account for only about 1 percent of the water on earth. Fresh waters, particularly the streams, are also a much more varied habitat. There are great differences in the oxygen content, amount of food, light, and currents from one body of water to another and also wide variations at different seasons in the same body of water.

STREAMS may be fast-flowing or slow and range in size from small and ditchlike to a mile or more wide. Some are cold, others warm. Some have a constant flow of water the year around; in others the flow varies from flood conditions during the rainy season to dry or nearly so at other times of the year. Cold, fast

Chinook salmon

Rainbow trout

Brook trout

Tadpole holds fast with suction disc

Brown trout

Salamander

Snail

Fast Fresh Water

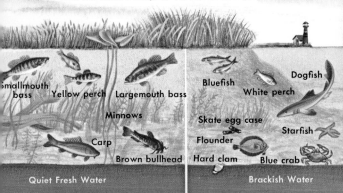

NATURAL WATER PURIFICATION IN A STREAM

streams are usually rich with oxygen, whereas slow streams have a low oxygen content.

Each of the different kinds of streams has a characteristic type of plant and animal life. Trout, for example, live in cold, fast streams, while carp are typical of warm, slow waters. Smaller animals that live in fast waters are either able to dart quickly from one protective rock to another or have flattened bodies or holding devices to prevent them from being swept away by the current. Tadpoles of fast streams, have suction discs on the underside of their body, as do some kinds of insect larvae. Some immature insects live in weighted cases made of small pebbles, twigs, or grains of sand glued together by a secretion from the insect's body.

Where streams are wider and slower, inhabitants resemble those of lakes. In these areas, the bottom is not washed clean by a current, for as the stream loses velocity, it begins to drop its load of silt or debris. The animals that live in these parts of the stream include many bottom feeders and burrowers.

Near oceans, streams may be brackish for many miles inland and may be affected by the daily tides.

Food chains in streams are linked closely with those of the surrounding land. Detritus, nutrients, and soil particles are fed into the stream from the watershed, hence the nature of the watershed really controls the quality of the water in the stream. The relationship of man to pollution of streams is obvious.

Illustration labels (top): unpolluted water · decomposition region · pollutants · degradation region · recovery region

Illustration labels (bottom): Smallmouth bass · Yellow perch · Largemouth bass · Minnows · Carp · Brown bullhead · Bluefish · White perch · Dogfish · Skate egg case · Flounder · Hard clam · Starfish · Blue crab · Quiet Fresh Water · Brackish Water

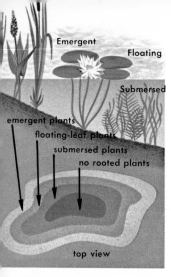

Emergent

Floating

Submersed

emergent plants
floating-leaf plants
submersed plants
no rooted plants

top view

Plants typically grow in concentric zones in shallow lakes.

LARGE LAKES are a more stable environment than are streams. They develop a characteristic population of plankton, both plant and animal types. The animals that live in large lakes may be divided into bottom dwellers, shore inhabitants, and open-water types, as in the sea. Algae, mosses, liverworts, and vascular plants grow along the shores and in the shallows. Close to shore, the plants may be permanently submerged but rooted in the bottom. Still closer to shore are plants rooted in the bottom and with floating leaves that shade out the growth of plants beneath them. Around the edges are rooted plants, their leaves and stems above the surface.

In deep lakes and ponds of temperate regions, the water near the surface circulates ver-

In deep lakes in temperate zones, complete "turnovers" occur in

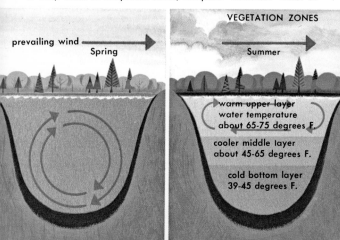

prevailing wind

Spring

VEGETATION ZONES

Summer

warm upper layer
water temperature
about 65-75 degrees F.

cooler middle layer
about 45-65 degrees F.

cold bottom layer
39-45 degrees F.

tically; the deep water remains stagnant, has little or no oxygen, and supports no aerobic life. Usually twice a year, these bodies of water "turn" completely, the stagnant bottom water coming to the top and the surface water going to the bottom. This regular "turning" distributes nutrients and oxygen throughout.

In addition to permanent standing waters that vary in size from the Great Lakes to half-acre ponds and smaller and from waters that are acid or are saltier than any of the seas, there are numerous marshes, swamps, and temporary bodies of water where the aquatic habitat begins to blend with the land. Some animals are so especially adapted to temporary ponds that they appear only during those brief times when water is available. The remainder of the year they are in a resting or inactive stage. Standing waters of this type are fairly common in the tropics and subtropics where there are distinct dry and rainy seasons. Some of the fishes, including lungfish and some kinds of catfish, can bury themselves in bottom muds and survive for many months until rains come again.

Each kind of lake must be analyzed to understand its unique natural cycles so that they can be preserved. With age, lakes become increasingly eutrophic—that is, enriched with nutrients. In the natural process, they will eventually fill and become land, supporting forests or other types of vegetation. Man's activities generally accelerate the process (p. 118).

spring and fall when water is about 39 degrees F. throughout.

IN A POND

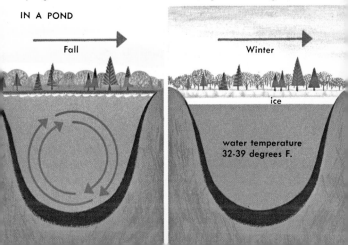

Fall

Winter

ice

water temperature 32-39 degrees F.

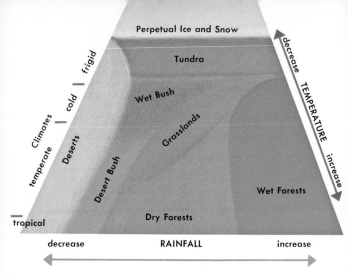

Perpetual Ice and Snow

Tundra

Wet Bush

Grasslands

Deserts

Desert Bush

Wet Forests

Dry Forests

Climates

tropical — temperate — cold — frigid

decrease — TEMPERATURE — increase

decrease — RAINFALL — increase

BIOMES

Terrestrial biomes are major ecosystems identified either by the plants that dominate the landscape, as in deciduous forests or grasslands, or by distinctive physical and climatic features, as in deserts and tundras. Biomes have developed over long periods of time through the complex interactions of soils, climate, and the organisms living in the region.

Each major biome is composed of numerous minor biomes or smaller communities (p. 63) that are formed as a result of different local conditions. The biomes generally described and shown on maps represent the vegetation that existed prior to man's destruction of large segments of natural vegetation over the earth. Man has left none of these areas untouched. Even such hostile regions as the polar areas and the deserts have been affected by man's disturbing activities.

SEAS are generally considered to represent a single biome, because of their uniformity. The common divisions recognized are described on pp. 46-49.

Freshwater habitats are seldom stable enough to be considered biomes, though some large lakes (p. 52) are similar to the seas in community structure.

INTERPLAY of two climatic factors—temperature and rainfall—largely determine the kind of ecosystem that occupies a geographical area. This is shown in the chart on p. 54.

To read the chart, consider the temperature to be horizontal, blending bands that represent the average temperature for the growing season. They range from slightly above 32 degrees F. (0 degrees C) at the top to 75.2 degrees F. (24 degrees C) across the bottom. This is the temperature spread from arctic or alpine conditions to the tropics.

Rainfall is represented as essentially vertical bands that range from a low (left) of about 5 inches (125 mm) to a high (right) of more than 300 inches (8000 mm) per year. Thus, the tropical rain forest is at the lower right, the tropical desert at the lower left.

In relation to latitude, the equator is at the bottom of the chart, and the polar regions are at the top.

SOIL (EDAPHIC) CONDITIONS are not reflected in the chart. Sandy soil, for example, does not hold moisture as well as does clay soil. An extensive area of sandy soil might support only a grassland, though the rainfall is enough for a forest region. In contrast, an underlying layer of hardpan that is impervious to water might prevent the downward percolation of water through the soil, thus making an area really wetter than is indicated by the amount of rainfall. The character of the vegetation would change accordingly.

TOPOGRAPHY is also a significant factor in ecosystems. In the Northern Hemisphere, south-facing slopes are much warmer than are slopes facing north. (The reverse is true in the Southern Hemisphere.) For this reason, physically adjacent slopes support completely different types of vegetation. In the tropics, high mountains may have ecosystems ranging from tropical rain forests at their base through elfin forests to alpine tundra at their peaks.

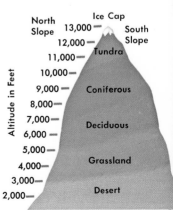

North Slope — Ice Cap — South Slope

13,000 —
12,000 — Tundra
11,000 —
10,000 —
9,000 — Coniferous
8,000 —
7,000 —
6,000 — Deciduous
5,000 —
4,000 — Grassland
3,000 —
2,000 — Desert

Altitude in Feet

THE ARCTIC TUNDRA

THE ARCTIC TUNDRA is a circumpolar biome north of the timberline and extending to the polar icecap. Its general character is a flat plain in which there are numerous small lakes. The surface of the soil around the lakes is rather uniformly fissured in a polygon pattern. Old lake sites are marked by mounds called pingos. At this high latitude, the growing season is short—about three months—and the ground is permanently frozen and non-productive except for a thin top layer that thaws during the brief summer.

LICHENS are the characteristic plant cover of the tundra. One of them is reindeer moss, a major producer in the food web. In the short summer, herbs must grow and produce seed quickly; the few woody species are dwarfed. Among them are willows, dogwood, bunchberry, and mountain cranberry.

TUNDRA ANIMALS include the caribou (reindeer), musk ox, lemmings, shrews, arctic hare, arctic fox, and wolf. Along the coasts are polar bears. Birds are abundant during the summer when many migratory species arrive to nest in the near-barren land. These include ptarmigans, owls, jaegers, sandpipers, and numerous waterfowl. Insects abound in summer, with swarming clouds of mosquitoes and black flies. There are also butterflies, bumblebees, collembolans (snow fleas), and other insects, all of which become inactive when it turns cold.

Food chains of tundra animals are short and have been investigated rather intensively, especially for radioactive pollutants. The lemmings, for example, feed on the plants of the tundra and then are preyed on by snowy owls, foxes, and wolves. When the lemmings reach a population peak, which occurs every three or four years, their predators also become numerous, lagging the lemmings by a season. When the lemming population goes back to a low level—actually "crashing" in a season—the predators also decline. They have no other choice of foods as they would in regions with a greater variety of prey animals. Some eat eggs, nestlings, or birds, available only in the summer months.

CONIFEROUS FORESTS occur as a broad, circumpolar belt of evergreens, the boreal forest, just south of the tundra in the Northern Hemisphere. The growing season lasts for three to six months; winters are extremely cold, with freezing temperatures at least half the year. Much of the world's lumber comes from this biome.

SPRUCE AND FIR are the principal trees in the northern portion of this biome, commonly referred to as the taiga. Farther south are pines and hemlocks. The shade from the dense canopy of the trees plus the thick carpet of needles that disintegrate very slowly results in a limited growth of shrubs and herbs.

Moose, black bear, wolf, lynx, otter, sable, marten, wolverine, squirrels, snowshoe hares, beaver, and numerous small rodents are among the mammals of the region. Birds include crossbills, grouse, siskins, jays, woodpeckers, nutcrackers, and numerous migrants in summer. Reptiles and amphibians are scarce because of the coolness, though a few kinds do occur. Insects are numerous in summer.

Because the forests consist of only a few species of trees, bark beetle, budworms, or other pests may quickly reach epidemic populations. They become a serious threat to the forests as well as upsetting food chains that must function rapidly so that animals can be prepared for winter.

THE DECIDUOUS FOREST BIOME occupies the temperate regions of Europe, Asia, and North America. It covers most of eastern United States, but as throughout the world, the original forests have been cut. Geologically, before the rise of the Rocky Mountains, this biome was probably a continuous belt across the North American continent. There is a limited deciduous forest region in South America.

MAN has flourished in this biome in the Northern Hemisphere. It is an area of four distinct seasons. The rainfall of 30 to 60 inches is fairly evenly distributed throughout the year; the summers are hot; the winters are cold but not severe.

This biome is considerably more complex than the coniferous because it consists of a number of climax vegetation types

that merge along the fringes with the coniferous forests. Common deciduous hardwoods are beech, maple, oak, hickory, walnut, buckeye, poplar, gum, magnolia, and sycamore. There are also many kinds of herbs and woody shrubs.

Similarly, the animal life of the deciduous forests is rich and varied. Black bear, deer, bobcat, gray and red fox, raccoon, mink, skunk, muskrat, cottontail rabbit, several species of squirrels, woodchuck, and numerous kinds of mice are among the mammals. Birds include many species of woodpeckers, hawks, owls, thrushes, vireos, warblers, flycatchers, jays, sparrows—most of them living in the region the year around but some migrating to warmer areas in winter. There are also large populations of reptiles and amphibians and many insects and other kinds of invertebrates.

Because man has occupied this biome intensively over many centuries, the character of both its fauna and flora has changed since Ice Age times.

Each species in deciduous forest develops a different autumn color.

Ostrich

Emu

Rhea

SOUTH AMERICA

AFRICA

AUSTRALIA

GRASSLANDS occur on all continents. Plains, steppes, savannas, prairies, pampas—these are all names for the same type of region where grasses are the dominant type of vegetation.

THE GREAT PLAINS—now largely under cultivation (the Grain Belt) or fenced for grazing cattle and sheep—represent the grassland biome in North America. Similar grasslands occur in temperate regions of America, Africa, Asia, and Europe; others in the subtropics and tropics.

In grasslands, the annual rainfall is 10 to 30 inches, too low for growth of forests but more than occurs in the deserts. The root systems of the grasses are typically much more extensive than the above-ground parts of the plants. During dry periods, the grasslands turn brown, but they green quickly with the coming of rains. Grass forms the base for the various food chains.

Large grazing mammals are the most conspicuous inhabitants of grasslands. On the plains of North America, the bison and the pronghorn were the dominant animals. Smaller mammals include the coyote, badger, kit fox, prairie dog, gopher, jack rabbit, and many kinds of mice. Prairie chickens, plovers, hawks, horned larks, meadowlarks, sparrows—these are among the typical birds. Reptiles are fairly abundant, but because of the dryness, amphibians are scarce. Insects are plentiful, with periodic epidemics of grasshoppers.

Most remarkable is the fact that the ecological niches of the grasslands on each continent are occupied with similarly adapted species. Thus the ostrich of Africa has the rhea as its counterpart in South America and the emu in Australia.

THE TROPICAL FOREST BIOME is located immediately north and south of the equator in Africa, Asia and many of the Pacific islands, South America, Central America, and the southern tip of the United States. It is most extensive in South America, where it covers nearly half the continent.

In ordinary usage, the tropical forest refers to the evergreen rain forests, where the rainfall may be 90 inches or more per year. The year is usually divided into a rainy and a relatively dry season, and the temperature is seldom lower than 70 degrees F. or higher than 90. There are also forests of deciduous trees where the rainfall is less, and in still drier regions, thorn forests develop. Mangrove swamps are a special community found along coasts. At the fringes of the forest region, it is sometimes difficult to determine whether it is a forest, a savanna, or a grassland.

VARIETY OF PLANTS AND ANIMALS is greatest in the tropical biome. A tropical rain forest may contain 500 or more species of trees, while a deciduous forest the same size consists of only a dozen or so species and a coniferous forest only two or three. The typical rain forest is composed of tall trees on which there are many vines (lianas) and epiphytic plants. The ground beneath the trees may be almost bare, with no undergrowth and with only limited amounts of plant debris because of rapid decomposition in the warm, humid environment. The trees are mostly broad-leaved, evergreen hardwoods, shedding their leaves continuously throughout the year, not seasonally.

Some animals are strictly ground dwellers; others live in the treetops where there are leaves in which to hide and both fruit and foliage to eat.

A first-time visitor to a tropical rain forest learns that he can see very little. During the day, most animals stay hidden in the high, thick canopy. The activity takes place mostly at night and at 80 to 100 feet above the ground. Then the tropical forest comes alive with noises.

Arboreal animals include rats, squirrels, mice, monkeys, sloths, and other mammals. Tree-dwelling snakes, lizards, frogs, and insects are abundant. Some species of ants and termites have abandoned the ground-dwelling habit and build nests in trees.

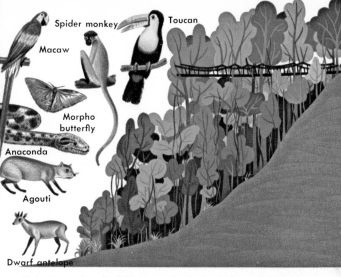

Biologists study animal life in top level of rain forest from bridge.

Birds are the most conspicuous animals in the tropical rain forests during the day. Parrots, parakeets, toucans, hornbills, cotingas, trogons, birds-of-paradise, hummingbirds—the colorful array is incredible and seemingly endless. Most of the birds are fruit and nectar eaters; some kinds eat insects. Birds of prey are less common in the subtropics and tropics than they are in temperate regions.

As in other biomes, the ecological equivalents on the different continents are striking. The agoutis of South America are matched in Africa by dwarf antelopes and in Asia by musk deer. Mammals of the tropics tend to be smaller than many of their relatives that live in temperate regions. This also includes the pigmy peoples that inhabit only tropical regions.

In contrast, coldblooded animals of the tropics are generally larger than their relatives in temperate regions. Here, for example, are found the giants among the reptiles—the big snakes, lizards, and crocodiles. The largest of the amphibians live in the tropics. Insects and other invertebrates also attain their largest size here.

Recent studies have shown that the tropical biome is extremely fragile. It cannot tolerate the agricultural practices that man has used in developing temperate regions. Unfortunately, such attempts are being made in tropical regions.

61

DESERTS are areas where the rainfall is less than 10 inches annually. The daily temperature fluctuation is great—from intensely cold at night to cool during the day or, in hot deserts, to as much as 100 degree F. or higher during the day. More moisture evaporates from a desert than is received in rainfall, and so the plants and animals of deserts are specialized to conserve water.

PLANTS are commonly widely spaced, which lessens the competition for water. Short-lived annuals bloom, seed, and die in a few weeks after a rain. Succulent plants store water and have shallow but widespread root systems that absorb water quickly. Others, such as salt-cedar, have deep root systems that reach to ground water. These plants use water lavishly.

Woody plants have tiny leaves on green branches. When the water supply becomes critical, the leaves drop off. This reduces the rate of evaporation. The green branches continue photosynthesis. Exposed surfaces of plants are covered by a heavy cuticle that reduces water loss. Desert animals use these succulents both for food and water.

Animals have an advantage over plants in being able to escape the heat in deserts. Most of them are active at night. In addition they have special adaptations to help them conserve water, such as nearly solid excretion rather than urine. Some can extract all the water they need from their food. Many are light in color and also have special flaps to close off their ears and nostrils to keep the sand out. They can either jump or run swiftly and may have fringes on the toes to help them get traction in the loose sand.

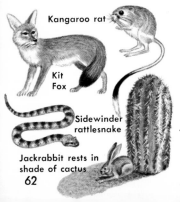

Kangaroo rat

Kit Fox

Sidewinder rattlesnake

Jackrabbit rests in shade of cactus

Some desert plants reach water table by means of deep roots; others store water in stems.

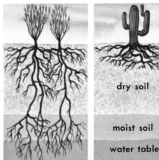

dry soil

moist soil

water table

Tree-island communities scattered in a grassland community.

COMMUNITIES

Communities consist of natural associations of plants, animals, and microorganisms that are directly or indirectly dependent on each other for their survival. These close-knit biotic units form the biomes. A ridge and a nearby low area can support a forest and a meadow community respectively. They are both part of a biome.

Communities must also be thought of in terms of time. The organisms living in a community modify and change the physical environment around them. Eventually, as a result of these changes, a dramatic shift in the population occurs. Communities are not stable, which is indicated by their natural succession (p. 82).

In some areas, the boundaries of the various communities are very sharp. This is especially true where the plant growth changes abruptly, as from trees to grasses within only a few feet. In others, the boundaries may be poorly defined, the different communities intermingling. These transition areas are called ecotones. If the ecotones are broad and have indistinct edges, the word "continuum" may be used to describe the condition.

63

brain   staghorn   elkhorn   star

Types of Coral

CORAL REEFS develop in warm, shallow seas (300 feet deep maximum, usually less) in a band around the world about 30 degrees north and south of the equator. They occur only where the temperature of the water is always 68-70 degrees F. or higher. The largest is Great Barrier Reef of Australia, more than 1,200 miles long.

REEFS are formed mostly of the limy skeletons of coral animals and by the calcified dead bodies of red and green algae and by the limy shells of various marine animals living in the community.

Reefs attached directly to land are called "fringing" reefs. Those separated from the shore by a channel or a lagoon are called "barrier" reefs. Where the sea rises over a mountaintop island, as it has in the Pacific, the coral may grow upward and form a doughnut-like ring around a central lagoon. This is called an "atoll."

The living coral animal is a tiny polyp, similar to a fresh-water hydra or to a sea anemone but encased in a limy cup. Great numbers of these cups are fused to form the reef structure. When the animals die, their stony skeletons remain, and new corals grow on top of them. Thus the reef expands outward and upward by a repetition of this process year after year. Only the top or outer layer of

the reef consists of living coral animals.

Some of the corals are flattened and crustlike in growth habit. Others grow in moundlike shapes, some are branching and fingerlike, and still others stand in tabular sheets. Altogether, the corals, sea fans, and other attached forms in their varied shapes and colors form a fantastic underwater "forest" that is inhabited by an equally incredible array of fish, shrimp, crabs, starfish, worms, and other animals. The coral reef provides an abundance of food and offers great protection in the numerous cracks and crevices as well as in the jungle-like growths. The wealth of life on a coral reef is unsurpassed anywhere else in the seas.

For many animals, the coral itself is food. Parrotfish have sharp, powerful beaks for crushing the limy skeletons to get at the soft insides. Butterflyfish use their slim beaks to probe into corals. Mollusks of all kinds in-

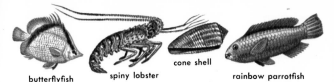

butterflyfish   spiny lobster   cone shell   rainbow parrotfish

Some Coral Reef Animals

habit coral reefs. These include both the pearl oysters and several kinds of edible oysters.

The most spectacular of the mollusks is *Tridacna*, the giant clam. It may reach a weight of 500 pounds—80 to 90 percent of the weight due to the heavy shell. The giant clam harbors algae in its mantle. During the day, the clam opens its shell and spreads its mantle in the sunlight in shallow water, and the algae carry on photosynthesis. The big clam gets a large share of its food from this "farm" in its shell.

The symbiotic algae in mollusks and coral play a significant role in the food chains of the reefs. They provide the basic food for various food chains. The plankton algae that perform this function in cooler waters are not abundant in warm tropical seas.

As durable as they may seem to be, coral reefs are nevertheless liable to destruction by both physical and natural forces. The greatest single threat in recent years in the Pacific has been the rapidly spreading crown-of-thorns starfish that preys on the living corals. This predator has now destroyed the reefs over many hundreds of miles along Pacific islands. Great Barrier Reef is now threatened.

A coral reef community in the Caribbean area.

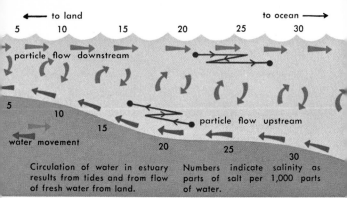

particle flow downstream

5

10

15

water movement

20

25

30

particle flow upstream

Circulation of water in estuary results from tides and from flow of fresh water from land.

Numbers indicate salinity as parts of salt per 1,000 parts of water.

ESTUARIES are near-shore waters where salt and fresh water are mixed and where the tides stir the water twice daily. Currents are generally strong in estuaries, due to the tides, the flow of a stream, or a combination of these two. The water is brackish—intermediate between salt and fresh—but the salt content varies greatly daily, seasonally, and vertically.

This dredge is pumping bottom sediments out of an estuary.

PLANTS, ANIMALS, AND MICROORGANISMS of estuaries are mostly marine species. Food is generally plentiful. It is brought in from the sea, produced locally, and also washed in from the surrounding land. In most estuaries, the bottom consists of a rich ooze of silt and organic matter. This results in a heavy growth of plankton that forms a broad base for estuarine chains and webs.

Estuaries are the nursery grounds for many species of fish and other marine animals. A large percentage of the fish that live on the continental shelves spend their early lives in estuaries. It is estimated that three-fourths of the commercial species of fish depend on their early life development in an estuary. Oysters and clams also live in estuaries, and it is in these waters that sea "farms" are located.

Unfortunately, estuaries are the portions of the sea being most steadily demolished by man. Along many coasts, a large portion of the estuaries have already been destroyed—by dredging and filling, by draining the salt marshes for insect control and to get more land for farming, by construction of highways and causeways that block the natural flow of water, by the digging of boat channels, and by pollution with sewage and chemical wastes or with choking quantities of silt. Estuaries are of vital importance to the seas. Upsetting these valuable communities can create serious problems in the total biosphere.

MUDFLATS are found along all shallow seashores. They may be a part of estuaries. Often the flats are exposed at low tide and flooded at high tide. Muds, composed of fine sediments mixed with organic debris, usually occur wherever the water is stagnated or protected from the action of waves and currents.

Mud contains food on which many worms, small crustaceans, one-celled animals, and bacteria flourish. Some kinds of mollusks burrow in the mud, actually eating their way through the organically rich deposits. Crabs are generally abundant on mudflats, and the detritus type of food chain is common.

Mud is low in oxygen, for much of the oxygen is used up in the process of decay. Hydrogen sulfide gas is prevalent. Both conditions limit the kinds of living things that can survive in the mudflat environment.

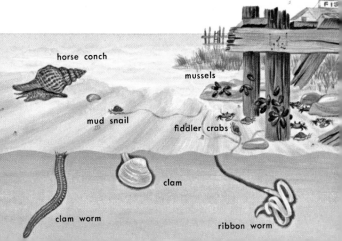

horse conch

mussels

mud snail

fiddler crabs

clam

clam worm

ribbon worm

hammock   pineland   palm-savanna

MANGROVE SWAMPS form a special community along seashores of the subtropics and tropics. Mangrove forests are composed of trees from three plant families in the Western Hemisphere, referred to as red, black, and white mangroves.

The red mangrove's stiltlike roots make it appear to be "walking" out to sea. Over a long period of time, this is literally true, for where the wave action is limited, the debris caught in the mesh of the mangrove roots slowly builds up the shore.

In addition, the mangroves will take root in shallow areas. They build islands that may in time join to the mainland.

The red mangrove's seeds germinate on the tree, producing cigar-shaped seedlings that drop and float. If they drift into shallow enough water, the tips become anchored, the roots grow, and the seedlings prosper.

Behind the red mangroves, often in places where the mud is covered with water only at high tide, are the black and the white mangroves.

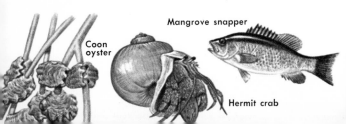

Coon oyster

Mangrove snapper

Hermit crab

marsh

white and black mangroves

red mangroves

offshore islands

Marine, freshwater, and land animals live in mangrove swamps. "Coon" or tree oysters grow on the arched roots of the red mangroves. Clams and a variety of snails peculiar to mangrove swamps may be abundant. Both fiddler and hermit crabs can be seen in large numbers.

Mangrove leaves, twigs, and bark are the base of a detritus food chain in the mud and shallow water. This chain is essential to the productivity of nearby estuaries and some ocean fisheries. The fall or organic matter is continuous. Large pieces are degraded piecemeal as they pass from one detritus feeder to another. Microorganisms finally finish the process. Detritus feeders are the prey of animals that move in from the sea, so productivity of the mangrove forest relates directly to the ocean.

Mangroves are a common hunting ground for raccoons, otters, and other animals, for food is plentiful. Some birds build nests and have roosts in mangrove swamps, their noisy rookeries filled with young birds.

Raccoon

Pelicans in rookery

SANDY SHORES are not as rich with plant and animal life as are many seashore areas. The sand is virtually sterile, supporting only sparse growths of plants along the beaches and fewer in the water.

WAVES, TIDES, AND CURRENTS shift the sand constantly, building it higher in some places and lowering it in others. Small storms continually alter sandy shores, and large storms may remove them. A sandy shore is highly unstable for life.

Nearly all of the forms of life on sandy shores live in burrows or hide in them when the tide is out. Parchment worms **dig** U-shaped burrows lined with a paper-like substance. Lugworms live in similarly shaped but unlined burrows. Plumed worms dig straight burrows and then extend themselves from the burrows to feed at high tide. Sand dollars, sea cucumbers, sand shrimp, flounders, stingrays, and many others squirm into the sand to cover themselves shallowly. Clams, conchs, cockles, and other mollusks burrow into the sand and leave only their breathing siphons above the surface. Nearly all of the burrowers can dig out of sight with astonishing speed—a race against the outgoing tide. Microorganisms also inhabit the upper layers.

Few animals live on the surface of the sand, but many kinds of animals come to the beaches to feed. As the tide moves in, fish follow it to prey on the sand inhabitants as they emerge in the deepening water. When the tide goes out, gulls, terns, and other shorebirds and even some mammals follow its wash to make meals of the sand-living creatures before they can scurry into hiding.

Shoreward from the high-tide line, a community of salt-tolerant plants grow. They help stabilize the sand, providing food and water for beach animals.

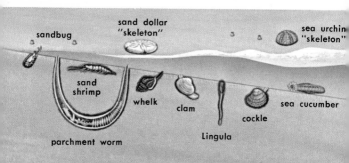

sandbug

sand dollar "skeleton"

sea urchin "skeleton"

sand shrimp

whelk

clam

cockle

sea cucumber

Lingula

parchment worm

ROCKY SHORES are inhabited by plants and animals that have developed strong holding devices or are able to hide in cracks and crevices to escape the waves. Some can bore into the rocks.

LIFE ZONES are clearly evident on most rocky shores. Some kinds of algae and crabs, snails, and other animals can live high on the rocks, their needs satisfied by the sprays and mists. Others can exist in a lower area that is submerged only at high tide. Still others are found only where they are always covered by water, even at low tide.

Among the animals adapted for life in the intertidal region are chitons, barnacles, limpets, mussels, oysters, and some kinds of sea anemones. A few seaweeds, such as rockweed and sea lettuce, grow here, but most are found only in the submerged zone. Here, too, are starfish, crabs, and other mobile animals.

Pockets in the rock kept filled with water by high tides are called tidal pools. They support special communities of plants and animals. Many pools contain permanent populations of seaweeds, fish, sea urchins, worms, shrimp, many small crustaceans, and other creatures.

Tidal pools are perilous places for life. In summer, such a pool can quickly become too hot for most living things to survive. During these times, too, the pool loses much of its oxygen, as it may also on cloudy days when

the plants continue to use oxygen but are not carrying on photosynthesis and thus replenishing the supply.

The salt content of a pool may rise too high if the days are hot and the pool is not flushed regularly by the tides. Or if there are many days of rain, the pool may become too low in salt content for marine plants and animals.

In subtropical regions, the water in the pools may become too cold in winter for the temperature-sensitive species. They cannot escape to deeper, warmer offshore waters as can species living in the open water.

Tidal pools develop permanent plant and animal populations.

OYSTERS were one of the earliest, if not the first, animals recognized to exist in communities. In a paper published in 1880, a European biologist wrote: "Every oyster bed . . . is a community of living beings, a collection of species, and a massing of individuals, which find everything necessary for their growth and continuance . . . Oyster-beds are richer in all kinds of animal life than any other portion of the sea-bottom." Even the oysters' shells become a habitat for other living things.

A VARIETY OF ANIMALS co-exist with oysters, and the oysters themselves can survive only in water that is rich with plankton. Drawn into the oyster shells with currents of water, the microscopic plankton are the food on which the oysters grow and fatten.

OYSTER-DRILLS, which are small snails, are among the predators that prosper where oysters are abundant. An oyster-drill uses its hard, toothlike tongue as a drill to rasp a hole in an oyster's shell. It works slowly and steadily, sometimes laboring for three weeks or longer before it can poke its toothy tongue inside the shell and lap up the juicy morsel. The snail then moves on to find another meal.

STARFISH are also common predators in the oyster community, and they can be a plague to commercial oystermen. Sometimes the starfish move onto an oyster bed by the thousands, each large starfish consuming as many as half a dozen oysters in a day. Within a short time they can destroy an entire crop.

LEOPARD RAYS also frequent oyster beds. They use their beak-like jaws to crack open the mollusks. Drum and sheepshead can also crush oyster shells. In addition, a variety of fungi, bacteria, and parasites may infest the oysters, sometimes reaching epidemic proportions. The oyster's shell is, in fact, a habitat for a remarkable assemblage of living things—sponges, sea anemones, protozoans, worms, barnacles, and many other creatures, all of them a part of the oyster community.

Some live inside the oysters' shells. Pea crabs, for example, make their home in the oyster's mantle cavity where they feed on the food strained through the gills. Several hundred of these tiny crabs may exist in only one oyster's shell, and in such abundance they do affect the oyster's ability to get enough food.

Only occasionally in the natural process do these various animals destroy an oyster bed, or if they do, the recovery is generally rapid. By commercial harvesting and by polluting and choking the water with silt, man

Oyster bar

Oyster-drill at work

Organisms that share oysters' shells

arfish attacking an ster

has in many cases permanently altered the habitat for oysters. Young oysters, or spat, need a hard surface to which they fasten themselves and establish or maintain a community.

Only a small percentage of the mass of shells in the oyster bed contains living animals. Over a long period of time, the shells and the organic debris that collects around it becomes a shoal. Eventually, it may be exposed for long enough periods to support plant growth. Islands can be built in this manner.

Oyster shells are barged back to reef to give young oysters firm places on which to grow. ▶

They may in time connect to the mainland and in this manner extend the shore as coral reefs or mangroves do.

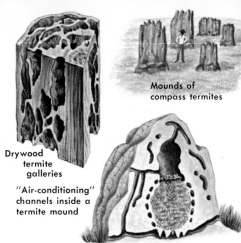

Mounds of
compass termites

Drywood
termite
galleries

Runways and nest
of tropical termites

"Air-conditioning"
channels inside a
termite mound

TERMITE COMMUNITIES are probably the oldest societies on earth, dating to at least 100 million years ago.
A community's survival depends on the ability of the
insects to protect their soft bodies from dehydration.
They accomplish this by living in dark, warm, damp
galleries, burrows, or carefully constructed nests that
are sealed most of the time.

Galleries and burrows are found in wood, soil and
up the foundations of buildings to connect soil and wood.
The galleries inside the wood insulate the colony from
the outside while at the same time the wood serves as
the food supply. In the wet tropics, covered runways are
built along tree trunks to connect a nest built in a tree
with the ground below. Nests of some tropical termites
are built around tree trunks and have shingle-like over-
laps, an adaptation to aid in shedding water during the
season of heavy rainfalls.

In the savannas, termites construct above-ground nests
by cementing together small pellets of soil. These mounds

74

are air conditioned by elaborate systems of tunnels connected to underground galleries. The well-known magnetic or compass termites build nests with their broad or flat sides facing east and west and the much thinner ends facing north and south. This is apparently done to get the full effect of the sun for drying out the wet clay brought from below the surface to build the nest.

FOOD CHAINS for termites are unique. Many kinds of termites eat wood. Protozoa that live symbiotically in their intestines secrete the digestive enzymes necessary to break down the cellulose of the wood, which the termites themselves cannot digest. Other termites—in more advanced groups—have evolved a symbiotic relationship with fungi that digest the wood.

In the tropics termites perform the same role that earthworms do in temperate regions. Their feeding habits and constant digging in the soil are important aids in keeping the essential elements (p. 24) in cycle. Termites thus have a great effect on the ecosystem in which they live.

OTHER INSECTS, unrelated to the termites, live exclusively in these communities. Some are strictly predators, some are parasites, and still others are simply tolerated as unobtrusive intruders. A few are invited into the community. A beetle that mimics a termite queen is welcomed by worker termites. They get fatty secretions from the beetle by grooming it in the same manner that they do the queen.

Aardvarks, pangolins, anteaters, and a number of kinds of birds prey on termites. The chimpanzee uses a straw to probe for termites in their burrows. Some anthropologists consider this to be one of the first uses of a tool by a primate.

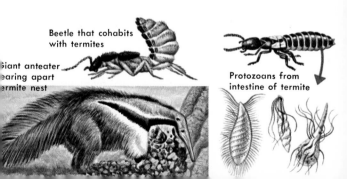

Beetle that cohabits with termites

Giant anteater tearing apart termite nest

Protozoans from intestine of termite

BOGS develop in lakes left by the Ice Age glaciers. They are common in the coniferous forest biome (p. 57). Because of their unique physical and chemical conditions that make them low in nutritional needs for nearly all organisms, these lakes are called dystrophic. As they fill slowly with organic matter, the lakes become bogs.

A high concentration of organic matter in the form of humic acids colors the water of dystrophic lakes brown. This limits the depth to which light can penetrate so that planktonic algae are not common. As a result, the oxygen level in the water is low. There is usually a deficiency of calcium, the water is highly acid, and the temperature of the water is low. All of these conditions combined reduce the process of decay, and so over a long time, undecayed organic matter accumulates on the bottom of the lake.

During the lake's stages of succession (p. 82) in becoming a bog, a floating mat of emergent vegetation starts at the shore, where it is anchored, and advances slowly into the water. The dominant plants in these mats are sphagnum moss and sedges. Organic matter accumulates abundantly under the mats, extending the shore. This process continues year after year, the lake becoming shallower and the amount of open water steadily decreasing (see illustration, p. 77). The rapidly filling lake is now known as a sphagnum bog.

Eventually the old waterlogged mats become nearly solid from top to bottom. At this stage, the old lake is referred to as a quaking bog. Another name is peat bog, because the partially decomposed plant material is peat—used as fuel and in plant nurseries.

The remaining areas of open water are surrounded by concentric rings of plants. From the middle outward, these are: (1) floating sedges and sphagnum and a few

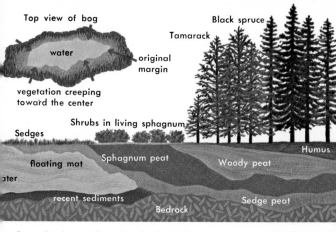

Top view of bog

water

original margin

vegetation creeping toward the center

Black spruce

Tamarack

Shrubs in living sphagnum

Sedges

Humus

floating mat

Sphagnum peat

Woody peat

ater

recent sediments

Sedge peat

Bedrock

bog shrubs, such as leatherleaf, bog rosemary, Labrador tea, and cranberry; (2) a dense stand of tamarack; and (3) a perimeter of black spruce. Insectivorous plants and orchids grow in the open areas.

Mosquitoes may be abundant in season, but animals are not common in bog water. Major food chains in a bog are based more on the already filled portions of the lake and the surrounding forest than on the aquatic environment. Insects support populations of amphibians that in turn are food for snakes and birds. Bears feed on the fruit of the shrubs. Shrews and rodents are abundant. Migratory birds use the bogs for nesting.

POLLEN PROFILES are made from studies of cores of peat taken from old, deep bogs. Because decay is retarded in a bog, the pollen grains are preserved as fossils where they fell and sank in the ancient lake.

Correlated with other ecological evidence, these studies are useful in learning the changes in climate that occurred during the many years of the Ice Ages. For example, if pollen from conifers is predominant in a core segment, years of cold weather are indicated—a glacial period. Pollen from deciduous trees, herbs, and grasses mark the long, warmer interglacial periods.

77

BLACK-TAILED PRAIRIE DOG COMMUNITIES can be seen today only in a small area in the southern section of the Great Plains region of North America. Before the settlement of the land, this was the most abundant animal in these grasslands (p. 59). Prarie dog "towns" seldom had fewer than a thousand animals, and one large community was estimated to sprawl over 30,000 square miles. These little animals are now rare and endangered, but the killing of them as "varmints" continues.

FAMILIES of prairie dogs set their own boundaries in their "towns," maintaining in this way a proper balance of numbers for the amount of food and the living space available.

A prairie dog rarely ventured more than a hundred feet from the opening to his burrow. Going farther was too dangerous, for out in the tall grass there were coyotes, wolves, and other predators. At a suggestion of danger, a sentry prairie dog on a mound would sound an alarm, and all of the prairie dogs would immediately scamper into their burrows. When the great herds of bison grazed through or near a town or came to wallow in the dust, they left the grass trampled or cropped. This helped the prairie dogs, for then they could see farther across the prairie.

Prairie falcon

Badger

Burrowing owl

The intricate maze of underground passages in a prairie dog town helped to aerate the soil of the plains. Burrowing owls lived in abandoned or little-used passageways. They took their share of the wandering young and also of the older, less alert animals. Black-footed ferrets preyed on prairie dogs regularly, as did also badgers. Prairie dogs were also victims of the prairie falcon, for the little rodents never learned to look up for the winged danger that swooped in silently with sharp, powerful talons to pluck them from their mounds. Surrounded by an abundance of food, the prairie dogs were plentiful. The natural predators did not dent their total population.

Man became the black-tailed prairie dog's greatest enemy. The rodents' burrows were a menace to horses, for they could easily break a leg by stepping into one of the holes. Further, both the mounds and the holes were not appreciated by plowmen, nor were the animals themselves when they ate garden and field crops.

Driven away by ranching and farming activities and shot or poisoned if they tried to stay, the vast numbers of prairie dogs disappeared like the bison. Only the slightly longer time and the less spectacular size of the animals detracted from the magnitude of the deed.

With these major links in their food chain gone, the black-footed ferret, prairie falcon, and burrowing owl are now rare and endangered, too. Badgers survived by moving into other regions, for their dependency on the prairie dogs was less direct.

Black-footed ferret

Black-tailed prairie dogs

MICRO-COMMUNITIES of many kinds exist within larger communities where habitat conditions differ greatly from those of the general area. These differences, though sometimes minute, include variations in temperature, humidity, winds, and similar physical factors.

DAILY AND SEASONAL CHANGES occur in some micro-communities. Above the forest floor, the humidity during the day is much lower than near the ground, but at night the humidity rises. Slugs, some kinds of insects, and other soft-bodied creatures may then venture off the ground and even high into the trees. They return to ground level during the day.

At some seasons, the forest floor becomes dry. Many of the animals then retreat into burrows below the surface to find the moisture level that is necessary for their comfort and survival. Desert animals that live in burrows actually create their own micro-climates in which the humidity is many times higher and the temperature much lower than at the surface.

A ROTTING LOG on the forest floor is a micro-community that may exist for many years. Ants, beetles, and termites cut channels into the soft, decaying wood that is being decomposed by rank growths of bacteria, molds, and other fungi. Lizards, salamanders, mice, and other small vertebrates use the rotting log for nesting and as a place of refuge. Eventually it crumbles and becomes an indistinguishable part of the forest floor.

Similarly, a micro-community may exist under a rock or beneath the bark of a tree. A hole in a tree is another place where a micro-community develops. The cavity perhaps served first as a nest for birds (owls, hawks, or woodpeckers) or for mammals (mice, squirrels, raccoons, or opossums). Some kinds of ants,

slug

snail

fungus

millipedes

white-footed mice

sowbugs

horn beetles and grubs

The bases of the leaves of epiphytic bromeliads that grow in trees in American tropics and subtropics collect rain water. These reservoirs serve the plants as sources of water and also become a habitat for tadpoles, immature insects, protozoans, and other small animals.

a bromeliad on the trunk of a tree

cut-away side view of reservoir with inhabitants

cross-section showing the numerous water reservoirs at bases of leaves

beetles, mites, and spiders move in quickly. Depending on the depth of the cavity and how much debris is accumulated on the bottom, a vine or other plant may get rooted there, the seed brought in by one of the occupants. Fungi grow in the dark, humid cavity.

Some species of mosquitoes breed only in the water trapped in holes in trees. In the tropics, there are frogs that depend exclusively on the water in these holes for the development of their tadpoles.

DEAD ANIMALS represent still another of the many micro-communities in which both the adult and immature stages of insects and particular kinds of bacteria are found. Some of the insects inhabit only carcasses that are still moist, feeding on the fleshy parts. Others specialize on the hair, skin, and similar dry parts, continuing to inhabit the remains long after the flesh eaters have moved to a new source of food.

Micro-communities exist also in the sea and in fresh water. Special groups of algae and marine invertebrates may be found only in sheltered bays or lagoons, on coral, or on pilings or other underwater objects. In these places the currents and salt content are different than in the sea generally. In fresh water, too, the down-current sides of rocks, debris in eddying currents—these or similar places may be vastly different than in the area generally.

81

SUCCESSION

Continuous changes, called succession, occur in a community over a period of time. Groupings of plants, animals, and microorganisms are replaced gradually by others until the complete character of the community is altered. Eventually a stable association of organisms is attained. This is the final stage, or climatic climax, in which the same kinds of plants and animals simply replace themselves. Successions have occurred in a continuing series since the first life appeared on earth.

Of the two basic types, primary succession starts in an area where life did not already exist. New islands, areas created by volcanoes, bare rocks exposed as a result of shifts in the earth's crust—these are the sorts of places where primary succession might be initiated. Secondary succession occurs when the normal stages of change are disrupted in an area but not all life has been destroyed, as after fires, floods, hurricanes, or agricultural operations.

The sequence of stages in succession follows a definite and orderly pattern, determined largely by modifications brought about in the environment by each plant and animal population as it paves the way for its successor. Rates vary, some kinds occurring rapidly and others requiring centuries.

POND SUCCESSION is a simple, easily understood example of primary succession.

In the beginning, the pond is only a water-holding basin, and in most cases, there is little life present. Microscopic plants and animals soon inhabit the water. Eggs of various kinds of aquatic animals and seeds of plants may be introduced to the pond on the feet of wading birds. Some insects arrive by flight, and other animals travel over the land to the water. Soon the pond supports a population of aquatic plants, fish, mollusks, and other creatures.

82

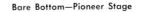

Bare Bottom—Pioneer Stage

Submerged and Emergent Vegetation Stage

Temporary Pond Stage

Forest Climax Stage

Plants begin to grow around the margin of the pond, which is then visited by a greater variety of birds and mammals. Organic matter builds up on the bottom of the pond and around the edges, and the plants begin to creep toward the center as the shore builds with organic re- mains and sometimes with sediments washed in from the surrounding land.

The character of the plant and animal population also changes —from strictly aquatic types to kinds that flourish in swampy areas. Trees and shrubs appear and, in time, a forest develops.

Ocean　　Beach　　Active Dunes

prevailing
winds

accumulated sand—washed up
by waves and blown by winds

DUNE SUCCESSION is another type of primary succession with clearly observable stages from beginning to end.

Dunes, commonly amassed along the shores of large lakes and sandy seacoasts, are bare sand in the first stage. Some dunes may be a hundred or more feet tall, consisting of loose sand shifted by the wind. Thus the dunes themselves are in motion. On the water side, where the sand is still accumulating, temporary populations of plants and animals occur. Carrion animals feed on the dead remains of plants and animals cast from the water. Still other animals prey on these scavengers.

Farther inland, the population assumes a more permanent character. The pioneer plants are kinds that can tolerate extremes of drying. They serve an important function in forming a halting barrier to the shifting sand, some

kinds even surviving burial by the sand, if at a moderate rate. Over many seasons the dead remains of these grasses and other plants are mixed with the sand, becoming humus.

A new community of plants now invades the progressively richer soil. They add more humus, paving the way for still another stage of succession. Soon willows, cottonwoods, and similar types of trees are growing in the old dunes area. All the while, the number of kinds of animals continues to increase, their wastes and their decayed bodies further enriching the soil.

In the Great Lakes region where the succession of dunes has been studied in detail, the climax stage is near when the old dunes support a pine forest. Soon after the pine forest stage, an oak-hickory forest will develop, and eventually a beech-maple forest.

Stabilized Dunes

humus, built from decaying plants and animals added to sand

Old Beach Surface

ABANDONED FIELDS are examples of secondary succession. One of the most studied of the abandoned fields involves the succession of old cotton fields in southeastern United States. Within 25 years, the old field has become a forest of young pines. In another 25 years the pines are mature but not reproducing. The pines are invaded by young deciduous trees that will eventually replace the pines. About a century is required for the land to reach its climax stage of a deciduous forest.

A badly eroded, old cotton field with abandoned farm buildings on its way to becoming a natural forest of loblolly pines.

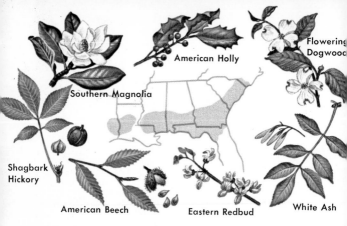

Flowering Dogwood

American Holly

Southern Magnolia

Shagbark Hickory

American Beech

Eastern Redbud

White Ash

SOUTHERN MIXED HARDWOOD FOREST is the climatic climax forest occupying the U. S. Coastal Plain from southern North Carolina to east Texas. (Hardwoods are broad-leaved trees as opposed to needle-leaved conifers, such as pines, which are softwoods.) This same region is part of the southeastern pine forest belt, the two forests intermingling. The pine forest represents the next to the final stage in succession, hardwoods replacing the pines on moister sites.

CO-DOMINANT SPECIES characterize this complex forest ecosystem. Common large trees are beeches, magnolias, oaks, hickories, sweet and sour gums, and American holly. Maples, basswoods, and ashes may also occur. Ironwoods, dogwoods, and redbuds are among the smaller understory trees.

Most temperate region forest climaxes are dominated by only two species and go by such names as beech-maple and oak-hickory climaxes. This forest, with its larger number of co-dominants, resembles the tropical forests because many species occur as scattered individuals. The magnolia is an evergreen broad-leaved tree, like many trees of the tropics.

The southern mixed hardwood forest represents a true climatic climax, for it is stable and self-duplicating. Minor climatic differences do not affect it, and when the forest is disturbed, it regularly returns as the same type in time.

SUBCLIMAXES sometimes occur within a climatic climax as a result of environmental factors. It is generally subclimaxes that produce the mosaic pattern of natural vegetation. These can be seen clearly in flights over a region or even when driving through a natural area. Variations in soil conditions and fire are two factors that commonly produce subclimaxes.

EDAPHIC SUBCLIMAXES are those brought about by variations in soil conditions, such as excessively wet or dry. In a southern mixed hardwood forest, for example, a low, swampy area will be covered with sour gums or cypresses. Trees that are not adapted to growth in soil that remains constantly wet cannot invade the area. Pines will grow on the very dry or sandy areas in the same forest region. These local forests add to the diversity of the area and increase the variety of niches that are available for animals and plants.

A cypress swamp of southern U.S.

FIRE SUBCLIMAX pine forests are common in areas that have a history of recurrent fires. Compared to broad-leaved hardwood seedlings that are easily killed by fires, pine seedlings and the mature pines are relatively fireproof. In such areas, grasses and a few herbaceous species take the place of broad-leaved woody plants, and pine woods become savanna-like. This condition is favorable for the lumbering and paper-pulp industries, but ecologically, these pine forests have much poorer food chains than do the pine-hardwood mixtures.

Controlled burning is a modern method of ecological control.

SEASONS are times of change that occur in all ecosystems from the tropics to the poles. They are most distinctive in temperate regions, which have spring, summer, fall, and winter. Even where seasonal changes are not so apparent, as in the tropics, in polar regions, and in high mountains, there are less dramatic changes that may nevertheless have a great influence on the living things in the communities. These changes help control the rate of succession.

Life in these communities must adjust to the seasonal changes that occur. Often there are complete shifts in population as well as in kinds of activities from one season to another. Any disruption in the normal rhythm creates a stress on the entire community. Low temperature can destroy flowers, thus eliminating the seed crop for the year. This affects the food supply of many kinds of animals.

CHANGES IN THE PHYSICAL or abiotic part of the ecosystem affect the organisms. These seasonal variations include the amount of light, temperature, precipitation, and atmospheric pressure. In polar regions, for example, the temperature may remain effectively stable all year, but the amount of light varies greatly, from winter days in which there is virtually no sunlight to summer when on some days the sun never sets. In the tropics, precipitation is the major factor in seasonal variations. Some tropical areas are divided into a wet season when heavy rains come day after day and a dry season when there is almost no precipitation for weeks. Most fresh-water communities, particularly those in shallow waters, are affected by these same physical variations. Marine communities, especially those in deeper water, are the least affected by seasonal variations, the greatest changes occurring in shallow water and in tidal zones.

CHANGES IN LIFE PATTERNS occur with the changes in the physical features of the area. In temperate regions, trees and shrubs lose their leaves in winter. The plants do not become totally inactive, however. Though they are no longer manufacturing much food, they are utilizing food resources stored during the warmer months of the year. Before their dormancy, buds form, and with the coming of warm spring days, the plants are ready to burst forth with leaves and flowers.

Beneath the plants, on the forest floor and sometimes covered with a heavy, insulating carpet of snow, small organisms continue to work through the winter, reducing the fallen leaves and twigs of countless seasons into organic matter that will in time be recycled into the soil and used again by the trees and shrubs. In the coldest weather, these organisms may move below the frost line in the soil. But their activity is unceasing.

Many birds and mammals continue their daily routines in winter but with changes in their diets. Animals that feed heavily on fruits, berries, and insects during the summer shift of necessity to nuts, roots, and mice or other small animals in winter.

They simply eat what is available. Most insects enter a stage of dormancy in winter, passing the cold months as inactive eggs or pupae. Other animals hibernate. In the dry season of the tropics and subtropics, some animals go into a similar state of torpor called estivation. Many plants also have seasonal periods of "rest" that are essential in their lives.

Some animals, most notably the birds, migrate from regions of extreme cold to places where the weather is warmer and food is easier to find. Some migrations are only down the slopes of mountains in winter, with a return to higher elevations again in summer; others are trips across continents and oceans.

Snow provides insulating blanket against drying winds and frigid weather. Beneath it life prevails through winter months.

Exotics are plants, animals, or microorganisms introduced by man to areas to which they are not native. Some introductions of exotics made are intentional because of a plant or animal's desirable features. Other introductions are made accidentally.

The ecological consequences are impossible to predict. Some exotics adjust to the new environment immediately. Often, in the absence of their customary competition for food and space and without the predators of their natural habitat, exotics undergo a "population explosion." Their numbers usually level off later as natural ecosystem checks and balances develop and begin to assert control. Other exotics may survive but in small, unnoticed numbers. Then, with changes in the environment, the population may suddenly expand. Still others cannot adapt at all and soon disappear. Some can survive only in the areas disturbed by man's activities.

Some successful exotics are welcomed, beneficial additions to the new habitat. As one example, the honeybee has been spread from its native Europe throughout the world. It has fit into the natural environment in a useful way as a pollinator of both cultivated and wild plants. Other successful exotics become pests, necessitating special efforts to eradicate them. The Norway rat is an exotic pest that man has taken with him wherever he has settled. Fleas, lice, house flies, and similar pests and parasites cohabit with man and his livestock. For many, their origins are now obscure. There are exotic plants, too, that serve man's needs in agriculture importantly. Others have become undesirable "pest" plants or weeds that have in some instances wrecked the natural ecosystems.

GIANT AFRICAN SNAIL, its shell 4 to 8 inches long, is raised for food in some parts of the world, but in most places is a pest. It forages on flowers, shrubs, and vegetable crops. Eradicating the snail is difficult.

INSECTS, because of their small size and also inactive and often inconspicuous resting stages, are commonly introduced accidentally. Shipments and baggage are customarily checked at ports to prevent hitchhiking pests and diseases from gaining entry. Many of the damaging pests in North America are natives of Europe—and vice versa.

The Colorado potato beetle lived originally in small numbers on the eastern slopes of the Rocky Mountains, where it ate a native plant of potato family. It did not become a pest until an exotic plant— the white potato from South America—arrived in its native range. Then the beetle began its trek eastward from potato field to potato field, also broadening its diet to include tomatoes, and other crops.

Insect pests have prospered as a result of modern agricultural practices that have made large continuous fields of their food available.

CHINESE MITTEN CRAB, its claws covered thickly with dark bristles, was an innocuous and accidental introduction to northern Europe. It now lives in the rivers and coastal bays along Baltic and North Seas.

BIOLOGICAL CONTROL has been effected by introducing exotic insects to kill undesirable species. A moth was taken to Australia from South America to bring under control the rampant growths of introduced cacti. Another classic example: the cottony cushion scale, threatening the California citrus industry, was brought under control by an Australian predatory lady beetle, the vedalia.

Vedalia

Cottony cushion scale

FISH introductions occur as a result of releases or escapes from aquariums or bait buckets (goldfish are an example of the latter) or the intentional introduction of a species by sportsmen and fisheries biologists.

SPREAD OF THE SEA LAMPREY through the Great Lakes was a classic example of what happens when an environmental barrier is removed. When the Welland Ship Canal was completed, the lamprey's path was no longer blocked by Niagara Falls. More than a century passed before the sea lamprey suddenly underwent a population explosion—in the 1930's.

Within a few years, the lake trout, sturgeon, whitefish, and other native species had been virtually eliminated by this voracious predator-parasite. Both overfishing and pollution of the lakes were also contributors. After much costly research, a selective poison was discovered that would kill the lamprey. Now the lakes are being successfully restocked with Coho and chinook salmon for sport fishing.

Many biologists are concerned that the proposed sea-level canal through the Isthmus of Panama might produce similar undesirable results. The now distinct and separated populations of the Atlantic and Pacific would be intermixed, and it is suggested that the total number of species would be cut approximately in half in the adjacent ocean areas.

CARP are exotics that have become very common in many lakes and streams in the United States. These Asiatic fish gained favor in Europe both as food and as wily fish to be caught on hook and line. In the late 1800's they were brought to the U.S.

Hardy, giant members of the minnow family, carp prosper almost everywhere they are introduced. In fact, they do too well. They root in the shallows, destroying nests of native species. At the same time, they spread their spawn copiously, each female laying as many as a million eggs in a season. In their favor, however, carp do thrive in waters that are too polluted for native fish.

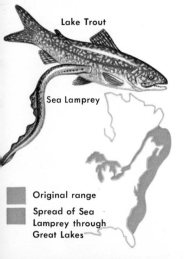

Lake Trout

Sea Lamprey

Original range

Spread of Sea Lamprey through Great Lakes

WALKING CATFISH were turned loose in a southern Florida canal by a fish dealer who had a surplus of the strange fish. Natives of Africa, the walking catfish can literally move across the land, slithering on their belly and using their stubby fins like legs. A labyrinth of chambers behind the gills trap air so that the fish can breathe out of water as though they have lungs. How to control them has confounded biologists. If poisons are put in the water, the catfish crawl out and hunt for a new location.

During a severe drought a few seasons ago, many bodies of water in the area dried completely. Others became stagnant. The walking catfish settled into moist mud, wriggling and squirming like lungfish to make cocoon-like enclosures in which they waited for the rains to come again. At the end of the drought, there were more walking catfish, spreading steadily. Unfortunately, the walking catfish competes severely with native fish, preying on them as well as usurping living space and food.

Walking Catfish

White Amur

BENEFICIAL FISH INTRODUCTIONS include the gambusias, or mosquitofish. These natives of southeastern United States are probably now the most widely spread species of fresh-water fish in the world. They have been introduced to warm-water areas around the world to help control mosquitoes. The little fish feed at the surface, consuming both mosquito larvae and pupae.

Brown trout, natives of central Europe, are stocked in cold waters throughout North America, while the North American rainbow trout has been taken to other continents. Largemouth bass are now caught in Hawaii, and several North American salmon live in New Zealand streams and coastal waters. The white amur is being experimented with in southern United States for the biological control of aquatic weeds. Often it requires study and observation over a long period of time to determine whether an introduced species will have undesirable effects on natural food chains or the habitat.

BIRDS, like fish, have become established both as escapees and as intentional introductions by sportsmen, wildlife biologists, and pet dealers. Many have been introduced by people longing for the familiar sight of a species from their homeland.

EXOTIC BIRDS escape regularly from tourist attractions. Some have become established locally. The striking blue-gray tanager of the American tropics is found now in the wild in the Miami, Florida area. More abundant and spreading is the red-whiskered bulbul, a handsome Asiatic bird that concerns many biologists because it is expected to find conditions suitable over a wide area and also because it feeds on fruit. Spotted-breasted orioles, mynahs, and a number of species of parrots and parakeets are among others now breeding in southern Florida.

ENGLISH SPARROWS are perhaps the most familiar example of bird immigrants in the United States, though starlings are also widespread. At the peaks of their population explosions, both of these birds have become almost intolerably abundant. Except locally, the populations of both have diminished.

Introduced from Europe in 1852, the English or House Sparrow spread rapidly through the areas shown in green on map.

English Sparrow

GAME BIRDS are introduced regularly by sportsmen and wildlife biologists. Because little study was given to the specific needs of the birds, many of the early attempts failed. Among the notable failures in the United States were attempts to establish the silver, Chinese, Reeves, golden, and other pheasants. Black grouse, button quail, and capercaillie were among the other failures. Gray and chukar partridges have done well, and the ringnecked pheasant is permanently established in the corn and wheat belts of the Midwest. Hunting pressure normally keeps the bird under control.

MAMMALS have been no less abundantly transported by man than have other animals. Generally, the effort has been made to introduce a desirable game species for hunting, but there are several notable examples of introductions for other reasons.

INDIAN MONGOOSES were imported into the islands of the Caribbean to control rats infesting the cane fields. They did the job well, eliminating the rats and at the same time increasing their own numbers. With the rats scarce, the mongooses turned to the native birds and other animals, exterminating virtually all of them. Efforts are still being made to get rid of the mongoose that has become a worse pest than the rats.

EUROPEAN RABBITS were taken to Australia and New Zealand, and in the absence of predators, the rabbits multiplied to a peak capacity. In a short time, they were destroying the range country. All types of controls have been employed—paying bounties, bringing in a virus disease (myxomatosis), and using poisons. The rabbits still persist. No other example shows as clearly what happens to a population when all competition and predation is eliminated.

MUSKRATS were introduced to Czechoslovakia from North America in 1927 and have spread over nearly all of Europe. Their population is estimated to be in the many millions. Meanwhile, in the muskrat's stronghold in southeastern United States, the giant South American coypu, or nutria, escaped captivity during a storm and is now spreading throughout the Mississippi drainage system. It threatens to take over the wetland niches of the native muskrat, particularly in the South.

From a few animals kept in a zoo in Czechoslovakia, the North American muskrat has spread over much of Europe and Asia since 1900.

Muskrat

PLANT EXOTICS fit both beneficial and harmful categories. An outstanding example of a successful introduction is the Douglas-fir, a timber tree of western North America that is now grown widely in Europe. Another is the Monterey pine, confined in its native range to a limited area along the California coast and not abundant. It has become an important timber tree in Australia where it was introduced. An unusual example, the ginkgo tree is planted widely in temperate regions as an ornamental but does not exist now in the wild in its original native range in Asia.

MELALEUCAS, trees native to the southern Pacific area, were introduced as ornamentals into southern Florida and California early in the 1900's. The cajeput, one species of *Melaleuca*, has invaded thousands of acres of fresh and brackish water wetlands as well as drier, sandy uplands. With their thick bark, cajeputs are virtually fireproof, and no serious pests or diseases have so far attacked the trees.

Cajeputs grow close together and crowd out other species.

AUSTRALIAN PINES (several species of the genus *Casuarina*) have also adapted readily to southern Florida at the expense of native species. Australian pines are not fireproof, however, and so their spread is checked to a degree by fires that occur in the dry season. They have spread over sandy beaches, where their root tangles prevent the loggerhead turtles from digging nests to lay their eggs.

Australian Pines produce deep litter, preventing undergrowth.

Lantana

Klamath Weed

Water Hyacinth

Prickly Pear Cactus

Elodea

LANTANA, a shrubby, sandpaper-leafed, prickly stemmed plant of the American tropics, is grown as an ornamental because its pretty blooms are produced nearly all year. Introduced to Hawaii, lantana grew rampantly over much of the grazing land. It was brought under control by introducing insects that ate the seeds. Lantana was taken also to Africa and is now a serious pest plant there.

KLAMATH WEED, a European plant, was first established in the United States near the Klamath River in northern California. Within a few years it spread over thousands of miles of range country. Given a choice, cattle would eat grass, but if the grass were browned, the cattle ate the green Klamath weed—and were poisoned. Reasonable control was finally achieved by introducing two related species of beetles from the weed's native habitat. The larvae eat the foliage, causing the weed to die.

PRICKLY PEAR CACTI taken from the American tropics to Australia as ornamentals soon spread over grazing land. Moths were introduced as controls.

WATER HYACINTHS, from South America, escaped into waterways of southeastern United States and now choke canals, bays and rivers with luxuriant growth. Eradication efforts are costly and unsuccessful.

ELODEA (*Anacharis*), an American waterweed, was taken to England accidentally with shipments of timber and quickly became a serious pest plant. It clogged rivers, prevented boating, stopped fishing with nets or with hook and line, and became so thick that swimming was impossible (with reported cases of drownings in the tangles). Before a means of control was arrived at, the waterweed declined on its own. It now inhabits the waters in a respectable state of balance and is not bothersome.

97

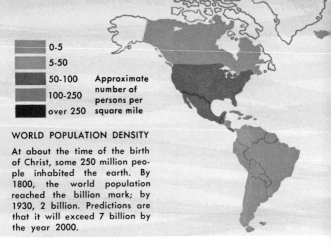

▬	0-5
▬	5-50
▬	50-100
▬	100-250
■	over 250

Approximate number of persons per square mile

WORLD POPULATION DENSITY

At about the time of the birth of Christ, some 250 million people inhabited the earth. By 1800, the world population reached the billion mark; by 1930, 2 billion. Predictions are that it will exceed 7 billion by the year 2000.

APPLIED ECOLOGY

Man's effect on the world environment has been so great that virtually no pristine conditions exist today, not even in the depths of the sea or in the upper atmosphere. For this reason, modern ecology is mostly "applied" (p. 7)—that is, working with these altered environmental conditions to achieve results that are desirable for all living things and particularly man. The efforts are to restore and sustain living conditions and to improve the quality wherever possible.

LAND USE has assumed major importance in recent years with the great increase in human population. Only slightly more than a fourth of the earth's surface consists of land areas. Less than a third of this limited area has been useful for livestock, crops, and living areas for man. Cold polar regions, dry deserts, rocky mountainous areas—these make up the remainder.

In years gone by, people used land until it was no

longer productive and then moved on to pioneer a new area. Except in some tropical areas, there are no more new areas to pioneer. The land must be used wisely so that it continues to be a livable habitat for plants, animals, and microorganisms, with basic natural cycles not greatly disturbed.

FOR MAN SPECIFICALLY, there must be adequate living space. Man is highly adaptable and can endure the most hostile climates, but living areas for large populations must be in reasonably hospitable environments. Ecologists are also concerned about the limits of crowding that man can tolerate. Large concentrations of people generate tremendous quantities of wastes that destroy the existing ecosystems not only for man but also for all living things.

Equally essential, land must be available to support livestock and to grow crops for feeding the increased population. Research must be continued to improve agricultural crops and techniques. Aquaculture, making use of both fresh and salt water for plant and animal crops, should supplement production on land, for a doubling of the world population within the next quarter of a century is expected.

Finally, recreation is one of man's subtle but basic needs to sustain his psychological well being. As an element of his innate being, man seems to need at least an occasional return to nature, of which he is a part by ancestral ties.

LAND RESTORATION may be necessary in badly misused areas. The rate of recovery depends on how badly the land had deteriorated and also on the use to which the land is to be put. Conversion to an entirely different kind of use is often wisest. In the South, the wornout cotton fields were in such poor condition, with little or no topsoil, that their return to agriculture within a forseeable time was impossible. Some of the areas were allowed to enter secondary succession (p. 82) and become pine forests again. In other cases they were put under strict forestry control for pine-tree farming.

PLANNING so that restoration is not necessary is one of the primary objectives of modern environmental sciences. Preservation does not mean that the land cannot be used. Rather, it is a way of preventing land abuse. A stream or a lake can produce large quantities of fish to be harvested for pleasure by sport fishermen. Not removing the excess population may, in fact, be harmful, for the fishermen represent a final and important step in the food pyramid. They prevent an overpopulation and a stunting of the fish in impounded waters. These waters can be clean as well as productive so that the domestic use of the water is not impaired even as they serve such recreational needs as boating, fishing, and swimming.

Ecologists examine aerial photographs with stereo-comparator (center).

STRIP MINING, in which coal or other resources are obtained from just under the surface, results in many acres of spoilbanks. The dug-out trenches become ugly scars, commonly filled with acid waters in which few living things can survive.

In the United States, for example, there are about 2 million acres of strip-mined land. Over the years a variety of ways have been tried to hasten these lands back into productivity. Some have been turned into tree farms. Where the condition of the water has permitted, the pits have been converted into fishing lakes to add to the recreation of the area.

Recently, thousands of these strip-mined acres have been converted into productive crop lands by covering over the gouged land with sludge, the end product of treated domestic sewage. The result has been a phenomenal growth of the crops planted as well as a valuable way of getting rid of sludge from a nearby metropolis.

Strip-mined land in restoration.

BEACH RESTORATION is now one of the major concerns of ocean engineering, but it is a problem only because of man's conflict with nature. The wearing down of coasts in some areas and their buildup in others is a natural process. Man has intervened by establishing, on his terms, an ownership of these coastal areas. Restoration is costly, and since the methods used still place man in competition with the oceans and other men, the victories are temporary.

Engineers halted the encroaching ocean (right) to protect the buildings on the beach and to extend recreational beach area (below).

SOIL CONSERVATION became one of the earliest concerns of ecologists in the United States because it involves the land on which man produces food. The need became apparent after the disastrous dust storms of the 1930's. It was obvious that the original farming methods exploited the land, causing a loss of valuable topsoil by both wind and water erosion. One soil conservationist estimated that an average of 8,000 acres of farm land was being carried to the sea every day by streams and rivers in the United States. This continued day after day, month after month, year after year—and still occurs, though to a lesser degree. In early farming, too, the same crop was planted year after year on the land until the soil was no longer productive and was abandoned. New land was then cleared for farming, until all of the available land was utilized.

Now recognizing the vital importance of retaining the soil and maintaining its fertility, agriculturists in most developed countries plow and plant on contours, alternate their crops, and plant in strips rather than in solid blocks. Productive soil is now recognized as one of the most valuable basic resources determining a nation's wealth and welfare.

MARGINAL LAND that did not produce has been removed from intensive farming and allowed to return to grass and woodlots. These vegetated areas help to reduce erosion and become good habitats for wildlife.

Farm ponds were built to help store water and to slow its runoff from the land. These ponds created a new kind of ecosystem and at the same time increased the wildlife population.

CHANNELIZATION, or the ditching of streams and rivers and the drainage of wetlands under the advisement of soil conservation agencies, has brought a negative reaction from most ecologists. Studies of channeled streams in several states have documented losses of as high as 99 percent of the stream productivity. Some streams and rivers are now scheduled to be returned to their original status.

Loose soils are easily transported long distances by winds.

EROSION occurs normally in nature. The mineral soil is built of eroded particles. But natural erosion is a slow process, often not removing the soil as rapidly as it is being built up. Man-induced erosion, by contrast, is very rapid and may result in sudden and dramatic changes.

The most important part of the soil is its uppermost layer—the topsoil. It is this portion, only a few inches thick in some regions but up to several feet thick in others, that contains the nutrients needed by plants for growth. Topsoil forms very slowly, yet all that nature has built over many centuries can be lost in a single downpour if the land lacks its protective cover of vegetation.

A single raindrop (right) may be the beginning of a deep gully.

Windbreaks separate fields and prevent erosion when fields are bare.

RENEWAL of worn-out soils and even restoration of badly eroded lands can be accomplished. If some of the topsoil still remains on the land, renewal may be reasonably rapid. The land is not row-cropped for several seasons but instead is limed, fertilized and planted in cover crops, such as alfalfa, clover, lespedeza, or some other legume. These may be mowed once or twice in a season to get hay for livestock. When finally plowed under, they add nitrogen to the soil in usable compounds as well as organic material to make soil humus. Wildlife also finds food and cover in these soil-enriching crops.

If the topsoil is completely lost, renewal comes much more slowly. It occurs by the natural process of secondary succession.

A dense vegetation windbreak can reduce a wind's speed by 50 percent.

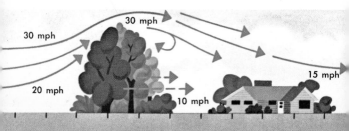

FARMING TECHNIQUES vary from place to place depending on climate, soil, and topography. But regardless of location, good farming protects the soil resource that was thousands of years in the making by natural processes in the ecosystem.

In areas of limited rainfall and of high winds, strip farming and windbreaks have proved to be useful protective procedures. In areas of heavier rainfall, soil erosion by water is a problem. Here contour farming can be added to the strip farming technique. The contours follow the natural elevations of the land, preventing rapid runoff and gullying. Water is thus kept on the land for use by the crops.

Strip-cropped land is protected from general wind erosion.

Contours of natural slopes are followed to prevent water erosion.

Modern single-crop farming has destroyed natural ecosystems.

AGRICULTURE, a top priority use of the land, represents a special kind of ecology, for to raise his crops and livestock, man eliminates the natural ecosystem and converts it to one of only a few species. Basically, then the farmer practices autecology (p. 7), working with the effects of soil, water, pests, diseases, and other factors on one or a few domestic plant and animal species.

To achieve results, crops must be planted at proper times and then carefully tended, for nature begins reclamation quickly. Yields are increased also by the development of new varieties that are most responsive to cultivation and also have resistance to pests and diseases. Chemical fertilizers are added to the soil to spur growth and to hasten the maturing of plants to harvest. Insecticides are utilized to kill pests; herbicides to stop the intrusion of weeds; fungicides or other chemicals to prevent the spread of plant diseases. All of these techniques have resulted in greatly increased production that, particularly in some of the less developed countries, has been heralded as a "green revolution." But ultimately, in all cases, natural ecological laws remain in control.

WASTES FROM AGRICULTURE are listed among the worst pollutants. Most of the chemicals put on the land are carried by runoff waters into streams and then into lakes or into the sea. Others appear as residues in food products.

As one of many examples, over half of the milk sold in the United States contains at least traces of insecticides. Many of these insecticides are stored in the body tissues of humans and other animals until a harmful dosage accumulates.

BIOLOGICAL MAGNIFICATION, a process that has put a number of animals on the endangered list, results from a small amount of insecticide in an animal's body being multiplied in the body of a predator as he eats more and more prey. In this way the top animals in a food pyramid suffer first and most severely. These include eagles, falcons, pelicans, and other birds that feed on smaller animals. As the chemicals build up in their systems, the females lay weak-shelled eggs.

Instances of direct kills are well documented, too. In using poisons to kill pests for health as well as for agricultural reasons, beneficial species are also destroyed. These include predators and parasites of the pests as well as honeybees and other insects that pollinate the crops. Even small amounts of these chemicals in water have caused major kills of fish.

Fertilizers have also caused major catastrophes. Used in prodigious tonnages, the excess washes into streams and eventually accumulates in ponds and lakes where it causes damage due to eutrophication (p. 118).

SAFER CONTROLS is a major objective of modern agricultural research. Most attention is given to biodegradable chemicals and to biological controls to replace insecticides and herbicides. These include the various naturally produced chemicals that are a part of the ecosystem in addition to encouraging natural predators. Insects are also being sterilized by radiation and by chemicals, causing significant reductions in their population.

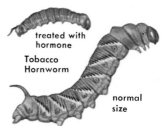

treated with hormone

Tobacco Hornworm

normal size

In biological control, a small amount of hormone interrupted normal growth of hornworm (above) and caused mealworm to become a misshapen undeveloped adult.

Mealworm undeveloped adult

FORESTRY is a special kind of land use, and there is increasing pressure for foresters to shift from their traditional emphasis on production and sale of timber to a greater focus on ecology and multiple use. In some cases, forests are maintained primarily to protect watersheds for streams, lakes, or reservoirs.

Selective cutting and planting, fire protection, controlled burning, control of pests—these are among the managment tools of the forester. As more research is done, the delicate balance of the forest communities is more appreciated. It is now more clearly understood that forests and their soils are highly sensitive to disturbances and that the ecosystem's productivity may be in jeopardy with improper management methods.

VIRGIN FORESTS of most of North America, Europe, and Asia have been cut. Second-growth forests now occur on some of these lands, either by natural succession or replanting. They are managed chiefly to produce timber and pulpwood. Clear-cut forest leaves land bare, subject to erosion. With selective cutting, a forest may be kept in constant production.

Natural growth of trees (left) compared to row-planted trees (right).

INCREASED AND SUSTAINED YIELD has been the primary objective of forest management in the past. Fertilization to replenish the depleted nutrients in the forest soils is now being given long-due attention.

Evidence indicates that forest soils are easily destroyed. Erosion is one of the factors involved. When forests are clearcut, the soil beneath the trees is exposed and may be swept away before vegetation can cover the land and prevent rapid and sometimes gullying runoffs.

Even if the soil appears to be retained, cutover lands are sometimes too deficient in nutrients to support new forest growth. The necessary elements and minerals have been leached from the soil, leaving it sterile or nearly so. This results in part from the change that takes place in the forest-floor communities after the cutting.

EXPERIMENTAL CUTTINGS in small watersheds show a great increase in the number of bacteria that convert organic nitrogen into soluble nitrates. These nitrates are carried off the land in solution in watershed streams, leaving the soil lacking in nitrogen. Careful and selective cutting appears to reduce the rate at which this takes place.

FOREST FIRES do not ordinarily damage the soil as much as clear cutting. The heat of the fire reduces the number of microorganisms in the upper layer of the soil, and the nitrate leaching (see above) is thus reduced. By the time this population is reestablished, the land is covered with a growth of herbaceous plants. There is evidence that the surface soil is made water repellent by organic material released in the fire. This may also help prevent nutrient leaching.

109

CITIES occupy only a small amount of the earth's total land area, but they are extremely complex in terms of human interactions, transportation, and communication necessary to supply their inhabitants with food and other materials. Totally, a third of the world's population is urbanized. Large amounts of land are needed to sustain them. For sound ecological reasons, cities must be linked to the land.

URBAN ECOLOGY deals primarily with cities, generally defined as places with populations of 100,000 or more. Cities have existed for more than 5,000 years, but until recent times they were relatively small and were supplied mainly by the land immediately surrounding them. The growth of large cities has come since 1900. All nations that are highly developed industrially are also largely urban.

Among the giants of the urbanized areas is the eastern seaboard of the United States, its complex of interconnected cities extending from Boston to Washington. More than 15 million people live in New York City and adjacent New Jersey, the heart of this sprawled metropolitan region.

As the population increases, the quality of the living goes down. Cities are thus the sites of countless socio-economic problems. Noise, air, and water pollution are serious problems. Transportation needs are great, for it is necessary for people to move to and from their work and also for goods and supplies to be moved to and through the cities. Automobiles and trucks are now recognized as major contributors to air pollution.

It is the cities, however, that have brought into sharp focus the great need for man to exist as a part of his natural world. Many solutions to ecological problems of cities suggested in recent years merely transfer the problem to other parts of the earth's ecosystem. The problems still exist and must be solved.

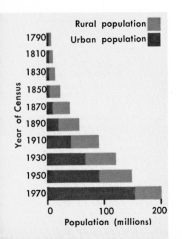

Growth of population has been most spectacular in cities.

WASTES, both industrial and domestic and as solids, liquids, and gases, are one of the major problems created by the concentrations of people in cities. Wastes are dumped into waters, released into the air, or spread on and under land. All dumping creates serious pollution that upsets the natural environment.

In recent years, an increasing percentage of wastes have been dumped into deep wells—half a mile deep or deeper. Initially, these wells seemed to be the safe and inexpensive way of getting rid of sewage and highly poisonous chemicals. Unfortunately, the wastes put into these wells do not always stay there. Leakage to the surface and into underground water supplies sometimes occurs. The wells must be abandoned or closely regulated.

Solid wastes (cans, bottles, paper, etc.) are no less of a problem. In modern societies, each person casts off above five pounds of solid wastes every day. Rarely are these wastes eliminated. Often they become scenic blights and health hazards. Incinerators are now outlawed in many areas because they pollute the air and still do not get rid of all the material.

RECYCLING of wastes is the wisest solution. This compares to nature's system where the elements and compounds are used again and again in the various cycles. A ton of paper that is returned to make paper again saves almost 20 trees that would have been cut for pulp. Junked automobiles can be reworked to get the usable metals and other materials. Similarly, cans and bottles can be reproduced to make new containers or used in other ways. Glass products, for example, can be pulverized and mixed with asphalt to make paving materials or with sand or other substances to make building blocks. Recycling has only begun to be explored.

Sanitary landfills are modern method of solid waste disposal.

final earth cover, 2 ft.

fence to catch blowing debris

bulldozer compacts wastes

scraper hauls cover material

original ground level

compacted solid waste

daily earth cover, 6 in.

WATER is the major component of protoplasm in both plants and animals and is essential in nearly all life processes, thus it plays a vital role in the biosphere. Over many millions of years, the amount of water on earth has remained about the same. Water covers roughly 72 percent of the earth's surface. In volume, there are about 1.5 billion cubic miles of water on earth, but 97 percent of this water is in the oceans. Only 3 percent is fresh water, and less than a third of this is usable. The remainder is locked in snow and ice at the polar caps and in glaciers.

In its natural cycle, water is returned continually to the atmosphere by evaporation from land areas and from the surfaces of oceans, lakes, and streams. Large amounts are also transpired by plants, passing up through the root system and then through the leaves into the atmosphere. This water vapor constitutes an average of about .03 percent of the atmosphere, but it is of great importance to life. It is by this constant

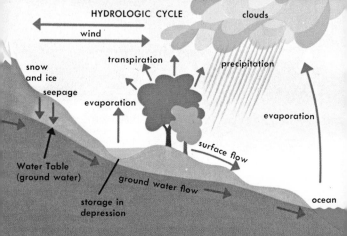

circulation that water which has run off the land is returned again to the atmosphere. It is this system too, that brings water back to the land from the oceans, after it is released from the atmosphere as precipitation —either rain or snow.

The precipitation does not fall evenly on the earth's surface. Rainfall is greatest in the equatorial regions and is least at the poles. There are also many local and regional variations. A cloud forest in Hawaii, for example, sometimes gets as much as 600 inches of rain per year; many desert regions, in contrast, receive less than 6 inches of rain per year. As rain-laden clouds rise over mountains or come over the land from the sea, they lose their moisture. Thus, the windward side of a mountain range is typically moist; the leeward side, dry. These are only a few examples of the differences in amounts of rainfall. Ecologically, the important factor is that the amount of rainfall determines the kinds of plants and animals that live in a region and also their density.

SPECIAL PROPERTIES of water make this simple, most abundant compound on earth distinctive. Water has a high specific heat. This means that it can absorb a large amount of heat energy without much change in its temperature. Similarly, it loses this heat slowly. For plants and animals that live in water, this is significant because temperature changes in their environment are not as rapid as on land. For the same reason, bodies of water moderate land temperatures. Land close to water remains warm longer in winter and stays cool longer in summer.

Water also has a great capacity to accept elements and compounds in solution. It is sometimes referred to as the "universal solvent." Nearly all of the elements and compounds needed by living things can be carried in solution. This is important not only for aquatic organisms but also for all living things, because the body fluids that transport food and oxygen and carry away wastes are principally water solutions.

At its freezing point, water is at a lower density than the water below it so ice floats. If this were not true, ponds, lakes, and streams would freeze solid from top to bottom in winter. But because ice floats, life can continue beneath winter's ice cover.

INCREASED USES OF WATER by modern societies have brought about water crises in many metropolitan areas in recent years. The total amount of water reaching the earth by precipitation has remained essentially the same and is adequate for sustaining all life on earth. Because of uneven distribution and poor management of the resource, water is regularly in short supply now in population centers where the need is greatest—for personal uses, for the production of materials, and for carrying away wastes.

DOMESTIC uses of water—for cooking, bathing, washing clothes, flushing toilets, watering lawns, etc.—average 75 to 100 gallons per day for each person. The minimum needed for survival by a human is only about a quart of water each day; all of the remaining gallons used simply reflect the modern way of life. Water is, in fact, the most used of all resources.

Supplying large cities with clean, drinkable water is becoming increasingly difficult. Reservoirs are quickly lowered in times of drought, even when the water is transported hundreds of miles. Cities that have grown where the water supply was never copious are turning to other means of getting fresh water. Near the sea, desalination plants convert sea water to fresh.

INDUSTRY consumes about as much water per capita as is used for domestic purposes. This doubles the total amount of water that must be made available in large population centers where industries are located.

Electrical generating plants—the power industry—utilize great quantities of water, too. They require totally as much as all other industries combined.

A few examples of industry's great use of water in production.

Listed below are a few common products and the water used in their production.

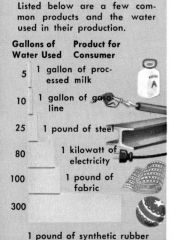

Gallons of Water Used	Product for Consumer
5	1 gallon of processed milk
10	1 gallon of gasoline
25	1 pound of steel
80	1 kilowatt of electricity
100	1 pound of fabric
300	

1 pound of synthetic rubber

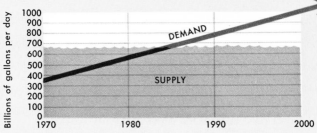

Shown above is the estimated increased demand for fresh water as opposed to the available supply, which remains constant.

AGRICULTURE has become an increasing user of water. Until recent years, water for irrigation was necessary only in the very dry regions, such as southwestern United States. But because precipitation that falls on the land now runs off more rapidly, with no cover of vegetation to act as a sponge to hold it back, and because of modern agricultural techniques that employ water and fertilizers to accelerate crop production, irrigation has become a regular practice even in regions where the rainfall is adequate. About half of all the water used in the United States now is for irrigation.

A portion of the increased use of water by agriculture can be attributed to changed patterns of living and also to the limited amounts of land available. In modern societies, diets have shifted from basic starches, such as rice and potatoes, to high-protein meats and to a variety of vegetables. All of these require greater amounts of water in production stages. Further,

this greater production has in many cases involved either increasing yields on land already in use or the turning to poorer lands that must be fertilized and watered generously.

RECREATIONAL AREAS, important to man's physical and mental well-being, usually include lakes, ponds, or streams. With the increased population and the greater amounts of leisure time, the need for more of these areas will grow. Boating, fishing, swimming, and similar kinds of recreation make a direct use of the water. Picnicking, camping, and other such activities may use the water only indirectly.

Recognizing the importance of recreation, modern planning includes the protection of these water resources for this purpose. Reservoirs for cities may be engineered to allow fishing and other recreational uses. Cleaned-up river basins can be attractive vacation areas while providing clean and usable water for domestic and industrial uses.

115

MISUSE OF WATER accounts for the shortages and for the lack of clean, usable water. The habit of the past was to use streams, lakes, and the sea as places to discharge all types of wastes. Because the wastes were fewer and also less poisonous, the water purified itself quickly. This is no longer possible, and the same water must be used again and again. Most impurities should never have entered the water in the first place.

POLLUTION of natural waters is a disgracing mark of modern civilization. The major pollutants, which deplete oxygen and at the same time add poisons to the water, are sewage from cities, industrial wastes, manures from domestic animals, and agricultural chemicals (fertilizers, fungicides, insecticides).

Streams, lakes, and even the giant oceans are now polluted throughout the world, worst near heavy concentrations of people. All of the major river systems in the United States are so badly polluted that they will soon be unable to support fish or any life of the original ecosystem.

New York City dumps more than 350 million gallons of raw sewage into the Hudson River every day. Sludges are barged a few miles offshore where the millions of tons dumped there move amoeba-

CURRENT SYSTEMS OF

None	**Primary**
1. Raw sewage collected and dumped into body of water.	1. Raw sewage collected, screened, and settled to remove solids.
2. Odor and disease problem acute.	2. Solids processed by burning, barging, use as fertilizer, etc.
3. B. O. D. (biological oxygen demand) very high.	3. Effluent released into body of water. Bad odor, disease problem, and speeded-up eutrophication still persist. Some solids still remain.
4. Speeds eutrophication.	4. Not as bad as no treatment.
Physical and biological processes, involving the bacteria of the natural detritus food chain plus the nitrogen, sulfur, phosphorous and carbon cycles.	

like across the ocean's now-sterile floor. The Potomac carries a similar load from the nation's capitol, contaminating the river with a hundred times more bacteria than the maximum allowable for safe swimming.

Cleanup is possible if started at the headwaters and continued to the mouths of streams. Treatments must involve multiple steps to assure clean water.

THERMAL POLLUTION became important with the advent of nuclear power plants that discharge excess heat into lakes, streams, bays, and harbors as electricity is generated. Most organisms are unable to live in water that remains above 85 degrees F. for prolonged periods. In many of the "heat waste" discharges, the temperature is consistently higher.

An estimated 20 percent of the total daily runoff water may soon be affected by thermal pollution. Cooling devices must be installed to reduce the heat of the water at discharge, or ways to convert the heat into a valuable resource must be discovered. Studies indicate that the heated water might be utilized to operate "hot house" vegetable farms that remain productive in winter as well as summer and also to maintain poultry, pig, and dairy farms. For people, the heat might be deployed to warm living quarters, to supply hot water, to operate air conditioners, and perhaps even to operate transportation systems. Some aquaculturists consider the heated water a boon because it permits the year-round culture of clams, oysters, and some kinds of fish.

SEWAGE TREATMENT

Secondary

1. Effluent from primary treatment is treated by trickling filter process or an activated sludge aeration system.
2. Solids are decreased but are still a problem as in the primary treatment.
3. Effluent has less impact on body of water but is still rich in nutrients, particularly nitrogen and phosphorous, and will speed eutrophication.
4. Not as bad as primary treatment.

Tertiary

1. Effluent from secondary treatment is treated chemically to remove most of the nitrogen and phosphorous, eliminating the eutrophication problem.
2. If chlorinated to kill disease organisms, effluent can be recycled into water system.
3. Conserves water
4. Best system but use is still limited.

Chemical processes where reclaiming and reuse of the chemicals is possible.

CHANGES OCCURRING WITH EUTROPHICATION

Biological

Increased growth of large aquatic plants, both floating and attached to the bottom.

"Blooms" of planktonic algae occur. Species of blue-green algae increase; other kinds decrease.

Floating and anchored algal mats appear.

Bacterial action increases. Fish and other animals sicken; fish kills occur.

Physical

Layers of dead algae and other plants accumulate on bottom, speeding the filling up of lake.

Water becomes distasteful and malodorous. Also changes in color, becoming red, green, yellow, purple, or murky.

Chemical

Dissolved oxygen drops from about 9 mg/l to 4 mg/l, at which stress on organisms occurs. Finally oxygen content drops to 2 mg/l, level at which death occurs. B. O. D. (biological oxygen demand) increases.

Nitrogen, phosphorous, and other elements increase.

Toxins increase.

EUTROPHICATION is a natural process in the aging of bodies of water. It means literally that the body of water is "well nourished," as opposed to its earlier oligotrophic age when it contained limited nutrients.

Small ponds may pass from the oligotrophic to the eutrophic stage in only a few years. The final stage is marked by a rapid growth of plants that choke out aquatic life and usher the pond from being a water environment to land. In large lakes, the natural process may require thousands of years.

The current interest in eutrophication has come about because of dramatic examples of man's acceleration of the natural process. Most specifically, Lake Erie reached a "near death" stage because its shallow waters were enriched with various pollutants—sewage, industrial wastes, and fertilizers. They made the lake super-rich with nitrogen and phosphates and brought about excessive growths of algae.

As these plants died and decomposed, they consumed the oxygen available in the lake, causing the death of other living things by suffocation.

A number of lakes have "died" in recent years as a result of this process. More are endangered. These bodies of water—most of them, at least—can be brought back to life by eliminating the pollution. Over a period of years, the waters will cleanse themselves so that living things can again survive in the habitat.

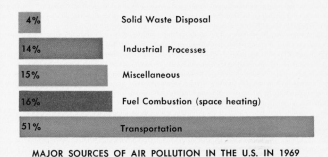

4%	Solid Waste Disposal
14%	Industrial Processes
15%	Miscellaneous
16%	Fuel Combustion (space heating)
51%	Transportation

MAJOR SOURCES OF AIR POLLUTION IN THE U.S. IN 1969

AIR forms the sea of gases, the atmosphere, that surrounds the earth. Close to the surface, its composition is remarkably constant—about 78 percent nitrogen, 20.9 percent oxygen, 0.03 percent carbon dioxide, 0.9 percent argon, and trace amounts of helium, krypton, and other rare gases. Air also contains water vapor, its amount varying from place to place and also at different seasons and times of day.

Air is heaviest (about 50 percent of its total weight) and contains the greatest concentrations of gases in a thin layer within about three miles of the surface. This is the troposphere, which is roughly 10 miles high at the poles and 5 miles at the equator. Other components of this envelope of air are the stratosphere (to 50 miles), ionosphere (to 200 miles), and exosphere (to 1,000 miles). Beyond these is outer space.

Living things are concerned almost wholly with the troposphere. They are dependent on the gases in this narrow belt. Animals, including man, can survive for relatively long periods without food and water, but they will die within a matter of minutes if deprived of oxygen. Pollution of the thin film of the atmosphere is thus of vital concern to man for himself and for all life.

119

AIR POLLUTION has reached global proportions. It is now universally obvious that only one sky covers the earth and that this sky—our atmosphere—has a limited capacity to dissipate wastes dumped into it. Pollutants that darken the sky at their source spread in a dimming haze for hundreds of miles. Large cities are capped by permanent clouds of dust and dirt that are so heavy they are little affected by winds and rain and so stable that they support permanent populations of bacteria and other organisms (p. 43).

In addition to these solids or "particulates," air pollutants include sulfur dioxide that originates mainly from the burning of fossil fuels, hydrocarbons from engine exhausts, oxides of nitrogen from power plants and automobile exhausts, plus ozone, carbon monoxide, lead, and other gases and solids that may be of local origin. In some areas, asbestos dust from construction has become a major air pollutant.

In many cities now, health officials advise against physical education programs in schools when the pollution index reaches a designated level. People with respiratory ailments are also told to avoid exertion at these times.

The total impact of air pollution on the biosphere is yet to be evaluated, but natural ecosystems are affected. Smog, for example, killed some forests of ponderosa pines in western United States.

TEMPERATURE INVERSIONS can increase the severity of air pollution. The temperature of the air usually becomes lower with increasing altitude, but occasionally a layer of cool air is trapped close to the earth with a warmer layer above it. This is an inversion. It prevents the normal rise of warm surface air to higher altitudes and thus acts as a trap to hold polluted air.

Deaths due directly to air pollution are easily documented in the case of inversions, such as occurred in Donora, Pennsylvania, in 1948, four years later in London, and still more recently in New York. In 1969, unusual climatic conditions held a heavy layer of pollution over a 22-state area of midwestern and southeastern United States for more than a week.

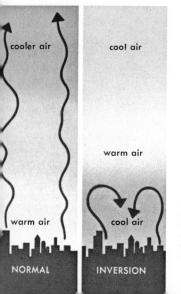

cooler air

cool air

warm air

warm air

cool air

NORMAL

INVERSION

Plants and mounds of earth properly placed act as noise buffers.

NOISE first became recognized as a health hazard in the 1960's. Attention was focused on noises caused by aircraft, particularly the sonic booms resulting when airplanes "broke the sound barrier." These abrupt, loud noises and the explosive impact of their sound waves do extensive damage to buildings, shattering windows, breaking plaster, and creating structural stresses. Many people complain also of nervous and emotional upsets, ulcers, heart diseases, fatigue, and general distressing annoyance. Manufacturers now produce airplanes with engine-muffling devices. Regulations prevent airplane take-offs over heavily populated areas.

Evidence indicates that these noises also affect wildlife. Mink ranchers, for example, report that their females will kill their young when the ranch is flown over by airplanes at critical times. Animals in the wild are no doubt similarly disturbed.

Almost every activity of so-called progress involves grinding, clattering, pounding noises that batter man's emotional makeup and thus indirectly cause physical harm. The normal sounds of any big city are loud enough to cause loss of hearing if a person is exposed to them day after day. Lulling noise-making devices that override outside noises are sold to help people go to sleep.

In various ways man is now insulating himself from the deafening and truly sickening sounds of his world. Industries are being redesigned in ways to eliminate noises. Garbage cans are equipped with padded lids and bases. Vehicles are checked for noisy mufflers. Mounds of earth as well as trees and shrubs are used as noise shields in landscaping. Air conditioning and soundproofing effectively block noises in buildings. These are a few of the ways man is reducing decibel dins.

Ecologists utilize the methods and knowledge of many different specialties (p. 6). This is essential because data from all of the physical and biological components of the ecosystems must be collected, stored, analyzed, and then interpreted in solving complex ecological problems.

A few well-established methods are described on the following pages. Procedures are revised constantly as improved tools or techniques are developed. Changes represent savings in time or increases in accuracy and sensitivity, coupled with modern techniques of automatic data and computer processing.

A SYSTEMS APPROACH TO THE STUDY OF A BADLY POLLUTED LAKE
—eutrophied due to sewage and agricultural wastes

Step 1
Measurement

Description of the problem and establishment of the objectives—specifically, the kinds of changes desired for revival of the lake. Included is a listing of the organisms wanted for the lake, the possible uses of the water, and similar kinds of factors. By field work, collections, and records of the data, this step also determines the current physical and biological condition of the lake.

Step 2
Data Analysis

Data from Step 1 is analyzed by computers, using statistical procedures to determine relationships between the various factors. As an example, the most important nutrients affecting fish and other aquatic life will be identified, establishing their source, at what times they are abundant or in critical supply.

SYSTEMS ECOLOGY is the ecologists' most modern and complex technique. It permits work with vast quantities of information to bring about an understanding of ecosystems and their intricate interactions.

Systems analysis resembles the time-tested scientific method of description of the problem, collection of information, theory, experimentation, and then testing for proof. In the systems method, great amounts of data are processed by computers in a much shorter time than was possible previously. Abstract, mathematical models of the real ecosystems are created, tested, and revised. These are then used as guides for research and also to make predictions about what might happen in the future.

Step 3
Modeling

Using the data analysis to select what appears to be the most significant relationships between the various physical and biological components of the lake, a mathematical model is created. The data from Step 2 become symbols and equations in a hypothetical explanation of how the lake ecosystem functions.

Step 4
Simulation

The model from Step 3 is simulated by computer runs. The various factors involved are assigned different values and then re-runs are made. In this way the effect of eliminating or changing factors such as the sewage effluent can be calculated.

Step 5
Optimization

Studies of several simulation runs are made to determine what solution is best. Costs must be considered in making a final decision, hence the course of action to be followed may be one of the alternate solutions rather than the one indicated as optimum by the calculations.

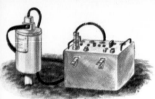

Neutron probe for measuring moisture at various soil depths over long periods of time.

Simple Berlese funnel for collecting small animals from soil or litter.

light—to create heat

soil or litter

alcohol

Plankton net for collecting small aquatic organisms

Secchi disc for determining water turbidity

DATA COLLECTING for analysis of an ecosystem requires equipment that can detect small amounts at high levels of accuracy. Even trace amounts may be significant, especially because of the possibility of biological magnification (p. 107). Temperature, light, pH, and similar factors may be difficult to measure in a meaningful way. Light and temperature, for example, commonly vary with the time of day, season, or even cloud cover. Devices that record constantly are most desirable but are expensive. Weather stations are usually established in or near the ecosystems being studied.

Special techniques may be required for collecting organisms. Many of these have been developed by specialists to give useful, reliable samples.

HISTORICAL INFORMATION often aids ecologists in understanding present-day ecosystems. Man has kept records only in recent centuries. Old weather records are useful, though they were kept reliably only in a few areas and offer limited kinds of information. Fortunately, there are techniques for interpreting natural records of much of the earth's history.

GEOLOGICAL RECORDS are useful in interpreting ancient habitats. The distribution of fossil species aided significantly in understanding continental drift as well as revealing much about how climates have changed. Ice Age changes in ecosystems have been traced by studies of fossil bog pollens.

The age of fossils is determined by chemical techniques such as carbon-14 analysis and others. By using O^{18}/O^{16} ratios in fossils from cores removed from the deep sea by drilling into the bottom, climatic curves for the past 200,000 years have been developed.

GROWTH RINGS in fish scales, clam shells, and tree trunks record environmental events and habitat conditions biologically. As a general rule, good growing conditions are indicated by larger and more widely separated rings. Trees usually produce one growth ring per year. Variation in the width or thickness of the rings reflect differences in the growing conditions.

Tree-ring chronologies of more than 7000 years have been made in southwestern United States. Their accuracy has been checked by carbon-14 dating. This computer-developed record is used to study past weather cycles and to establish reasons for the migration of early man into and out of the region.

Tree rings of a living tree can be sampled with an increment borer causing little injury. Charcoal from old ruins is also studied—for ring information.

Core removed from still-living, 400-year-old tree

Core from nearby stump shows that tree lived from 1200-1690, overlapping living tree above

Third core from a timber in an early cliff dwelling was from a tree 200 years old when it was cut in 1250

Rings are studied and counted carefully, often with the use of carbon-14 to establish correct dates.

1972
1872
1772
1690
1572
1490
1250
1200
1150
1050

MONITORING is the systematic detection and measurement of the physical, chemical, and biological components that interact in an ecosystem. Coupled with application of knowledge gained from systems ecology (p. 123), monitoring is the key to maintaining a quality environment for life on earth.

THREE BASIC SERVICES can be accomplished by monitoring: (1) establishment of the present status as a base line for environmental conditions; (2) warning of critical changes; and (3) assembly of measurements and data that can be used in continued management.

Pioneered by large cities, monitoring systems now have global applications. The 1972 United Nations Conference on the Environment asked for establishment of early warning systems for contaminants of water, air, and food and for programs to prevent the crossing of international borders by pollutants. This can be accomplished only by monitoring.

Monitoring procedures and the data collected should be uniform throughout the world so that the information is useful on a global basis. Environmental problems are now of a magnitude affecting all peoples. Common efforts on the part of large political powers in recently shared studies of the oceans, in world-wide health programs, and in earth-orbiting satellites have helped greatly in unifying techniques and in arriving at solutions to common problems.

SOME ATMOSPHERIC POLLUTANTS CURRENTLY BEING MONITORED
(by the National Air Pollution Control Administration)

Gases		Radicals
Carbon monoxide	Boron	Ammonium
Methane	Cadmium	Fluoride
Nitric oxide	Chromium	Nitrate
Nitrogen dioxide	Cobalt	Sulfate
Pesticides	Copper	
Sulfur dioxide	Iron	**Others**
Hydrocarbons	Lead	Aeroallergens
	Manganese	Asbestos
Elements	Mercury	Benzene-soluble organic
	Molybdenum	compounds
Arsenic	Nickel	Pesticides
Barium	Tin	Radioactivity
Beryllium	Titanium	Respirable particles
Bismuth	Zinc	Total suspended particles

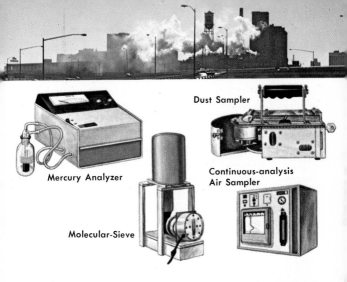

Dust Sampler

Mercury Analyzer

Continuous-analysis Air Sampler

Molecular-Sieve

MINUTE AMOUNTS of materials must be identified and measured by monitoring systems. In everyday measurements, an ounce is considered to be a small amount. In actual ecosystem measurements, which are usually done in the metric system, an ounce is about 28.3 grams. Liquids are measured in liters, which equal approximately a quart. Pollutants are usually so potent biologically that much smaller units must be used.

Parts per million (PPM) is a common system of expressing the amount of a pollutant. If one gram (1/28.3 of an ounce) is divided 1000 times, the amount is a milligram. If this amount of a substance is put into a liter of liquid, it becomes one part per million (PPM) in the liquid. Some substances are so active that

even smaller amounts are significant. For these, parts per billion (PPB) is the measurement. A milligram divided 1000 times and one of these units added to a liter of liquid is one PPB. Similar low levels are measured in monitoring radiation. The units of measurement are millirads and microcuries.

Laboratory tests and field measurements have shown that some insecticides can kill fish if present in the range of 0.1 to 0.01 PPM. Blue crabs may be killed in about a week's exposure to one PPB or shrimp in about two days of exposure to 0.3 PPB of some chemical poisons. Fish containing more than 5 PPM of DDT cannot be sold for food. The average concentration of DDT in coho salmon taken from the Great Lakes is 19 PPM.

REMOTE SENSING is the gathering of biological, chemical, and physical information in an ecosystem by the use of distant sensors. These may be cameras carried in aircraft or in spacecraft, or pictures may be taken from platforms or from the ground. With remote sensors, large amounts of data can be collected in a short time. Inaccessible locations may be studied remotely. These information-gathering devices can make it possible to solve problems that would be difficult or impossible with traditional techniques of autecology and synecology.

Sensor systems in use are: aerial photography, using black and white, color, and infrared films; radar; laser; a complicated system known as multispectral scanning;

and thermal scanning. All of these make use of the electromagnetic spectrum. Some systems are adapted to computers and densiometers to increase effectiveness.

Remote sensing is possible because each kind of object or organism receives and then reflects or absorbs electromagnetic energy differently. As a result each registers differently on film. For example healthy green leaves show green in color film, as tones of gray in black and white, and shades of red in infra-red. These identifying characteristics are called the object's signature. This may vary with the time of day, with the season, and with the physiological condition of the organisms being photographed.

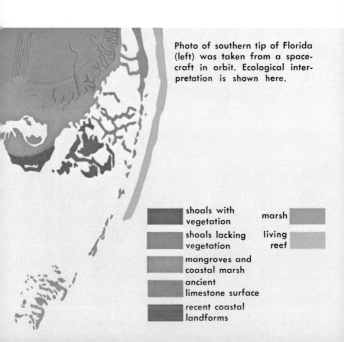

Photo of southern tip of Florida (left) was taken from a spacecraft in orbit. Ecological interpretation is shown here.

shoals with vegetation

shoals lacking vegetation

mangroves and coastal marsh

ancient limestone surface

recent coastal landforms

marsh

living reef

AERIAL PHOTOGRAPHS most useful for ecosystem studies are taken directly over the target rather than to one side or obliquely. For exact analysis, the scale of the photography must be accurate, possible only when the altitude of the camera is precisely known or when objects of known size are included in the area.

Much more can be learned from photographs taken with stereoscopic coverage. A two-lens camera system, one lens located ahead of the other, is used, operated so that the exposures have 60 percent overlap. The photographs can be used singly, with or without magnification, or in stereoscopic pairs.

Adjacent photographs make a stereoscopic pair that can be viewed and analyzed with a stereo-comparator. The area then appears in three dimensions.

Recognition of objects is easy, and even differences in the height of objects can be measured. Accurate maps can be constructed. The standard stereoscopic pairs of photographic prints are 9 x 9 inches. Large scale (70 millimeter) photography is used to make stereoscopic photographs at a very low altitude. These give more detail to supplement high altitude coverage. Tree species can be identified, forests inventoried for the condition of the trees and for yields of lumber, and even some kinds of insect damage or disease can be seen.

Helicopters operating at low speed and altitude are ideal to get large-scale photographs of spot locations. Ground markers are used to give an accurate scale, and it is also possible to take stereoscopic photographs from helicopters.

Infra-red aerial photograph shows pine-beetle damaged trees yellow. The uninfected trees are red.

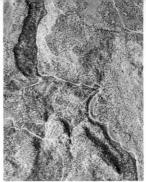

Stereoscopic pair of aerial photographs (above) become three-dimensional when viewed with stereo-comparator (p. 100).

Black-and-white photograph at left was taken with panchromatic film; photograph at right was taken with infrared film.

Side-look radar image of San Diego Harbor shows details of city and kelp beds (upper left) invisible to ordinary camera.

RADAR has an advantage over other remote-sensor systems because it can be used at night and also when weather conditions prevent the use of cameras. Side-looking radar produces a continuous strip image that can be recorded on film. A further advantage of radar is its ability to penetrate vegetational cover so that ground-surface features and even below-ground features can be sensed. Radar returns that produce light tones on the photographs usually indicate man-made objects; medium-gray tones, vegetation; and black, water. Radar is used also to determine temperature and moisture conditions. The sensitivity is so great that a single flying honeybee can be tracked.

Radar picture of Panama's Darien Province, which is always too covered with clouds to be photographed by usual means.

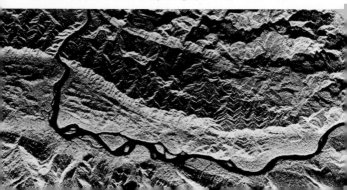

| Aerial Photograph | Computer Printout of Vegetation, Soil, Water | Computer Printout of Water Only |

By use of computer, special aspects of terrain can be separated.

MULTIBAND SPECTRAL RECONNAISSANCE, also called multispectral scanning, is a complicated system involving aircraft, sensors, spectrophotometers, and computers. As many as nine lenses may be used simultaneously. The reflected electromagnetic energy is sensed and split into several channels by wave lengths. Each spectral band (wave length) is recorded on a tape for computer printout.

This form of remote sensing can be used to extract much more information about the target area than with any other system. Its greatest current use is in surveying agricultural ecosystems. Crops can be identified and judged for health and yield. The printouts of farm areas show fields of oats designated by the letter "O"; wheat is indicated by the letter "W," etc. To date the system has not been applied to natural ecosystems extensively.

GROUND TRUTH is a necessary part of remote sensing. The area must be visited either before or after the remote sensing. In this way, the exact source of signatures can be verified. Samples of vegetation can be taken, and presence of animals on the survey tract noted.

Exact measurements are made so that the scale of the photographs can be verified. When this is done for part of the target area covered in a stereoscopic pair, the entire area can be interpreted with accuracy.

133

Analysis of stomach contents of predators or the regurgitated pellets of owls or similar predators are used to determine food habits.

FOOD CHAIN ANALYSIS

FOOD CHAIN ANALYSIS is essential in studying an ecosystem. The number of trophic levels should be determined, the organisms identified and counted, and the amount of energy flowing through the system measured. Obtaining this information is both difficult and time-consuming, and it is complicated by the fact that ecosystems change constantly. The methods used vary with the habitat, the kind of equipment, the time available, and the degree of accuracy necessary.

PRIMARY PRODUCTIVITY is the basic measurement, since the consumer part of the food chain depends on this basic production of food. It can be measured by several procedures, such as (1) weight of the plant material produced, (2) amount of carbon dioxide used by photosynthesis, (3) amount of oxygen released during photosynthesis, and (4) amount of chlorophyll present for photosynthesis.

Of these four procedures, the last three require elaborate equipment and also demand experience in working with data of this sort.

WEIGHING plant material is the easiest way to measure productivity. The harvests are taken from measured plots, called quadrats. If several different consumers feed selectively on different kinds of plants in the quadrat, the harvesting may be done species by species.

The plant material is weighed, dried, then weighed again. This gives the fresh or wet weight and the dry weight (hence the water content). To determine energy content, weighed dry samples are burned in a calorimeter (p. 19) and the results expressed in Kilogram calories (K. cal.).

CARBON DIOXIDE measurements are made in terrestrial ecosystems to determine productivity indirectly. A transparent chamber is placed over the plants so that photosynthesis continues, using carbon dioxide. The amount of carbon dioxide that exists in the incoming air as opposed to the outgoing air can be measured by air sampling. The difference, after an adjustment is made for the carbon dioxide lost in respiration, represents the amount used in food production. Carbon dioxide lost in respiration is measured either by using paired dark and light chambers or by making the measurement at night.

The variability of the entire terrestrial system being studied must be considered in making these measurements. Separate chambers may be necessary to get samplings of different species. If enclosures are not necessary, tubes may be used to suck in the air at a fixed rate from various locations in the ecosystem. Samples are collected from each tube for analysis, and in automated setups, the measurements are recorded on a continuous drum recorder and later analyzed by computer.

OXYGEN production by photosynthesis can be measured as a means of establishing productivity in an aquatic ecosystem. The amount of dissolved oxygen (D. O.) is determined chemically by the Winkler method or electronically by oxygen electrode.

The Winkler method depends on fixing the oxygen with chemicals in such a way that iodine (representing oxygen) is released and measured. The solution used is so accurately made that when a 200 ml sample of water is tested, the dissolved oxygen, as parts per million, equals the milliliters of solution used in the measurement.

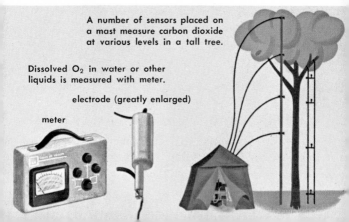

A number of sensors placed on a mast measure carbon dioxide at various levels in a tall tree.

Dissolved O_2 in water or other liquids is measured with meter.

electrode (greatly enlarged)

meter

POPULATIONS are difficult to study. Many techniques have been developed for measuring and evaluating them. Improvements in the methods are made constantly. Usually faced with time limitations, ecologists rarely attempt a total census. Rather, they try to get a reliable estimate of the population by measuring samples. No matter how large, variable, or unevenly distributed the population is, the size and number of these samples must be statistically representative of the entire population. Some of the methods for measuring animal populations are described on p. 142.

QUADRATS are common sample units for population measurements. A temporary quadrat is used only once. Permanent quadrats are selected carefully and are marked so that they can be found and studied for changes through periods of time. The size and distribution of all the plants and/or animals in a quadrat are listed. Habitat descriptions are often supplemented with photographs. To study the effects of animals on the plants in an ecosystem, animals may be kept out of the quadrat; this is called an enclosure—in contrast to an exclosure quadrat.

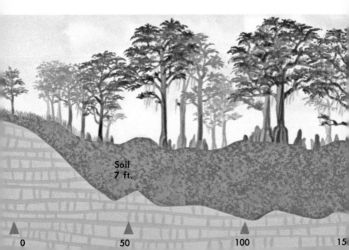

Soil
7 ft.

0   50   100   15

SHAPES AND SIZES of quadrats vary depending on the objectives and on the kind of population being studied. Originally, all quadrats were squares, but it has been demonstrated that data varies less when the quadrats are rectangular. Quadrat size is determined by the size and number of organisms being measured. Quadrats for trees may be 30 x 30 feet; for shrubs and small trees, 12 x 12 feet; and for herbs, 3 x 3 feet. Mosses and small insects may be studied in quadrats of only 4 x 4 inches.

NUMBER AND LOCATION of quadrats is determined after inspection of the entire area to be studied. The problem is most complex in large, variable areas. The quadrats may be located at random, equally spaced, or sometimes along compass lines.

TRANSECTS are long, narrow sample units that may vary from several miles to only a few feet in length. The size of a transect is determined in the same way as is the size of a quadrat. Transects are particularly useful when the objective is to learn how individuals are distributed within and between several adjacent communities. Line transects are made by stretching a tape and recording the species intercepted by the tape.

BISECTS are transects that sample the vertical and linear distribution of plants and animals. They show the layers of the habitat in profile. Some bisects include the distribution of root systems and soil animals as revealed by a ditch dug along the transect line. Often the plants are illustrated by symbols.

Limestone

This is a detailed transect of a cypress forest showing depth of soil built over limestone bedrock. Deepest soil is 7 feet; the tallest tree is 90 feet.

200   250   300 feet

WILDLIFE MANAGEMENT

Management of wildlife, which includes fish as well as game birds and animals, is a broad and important area of applied ecology. Because of the popularity of hunting and fishing, funds were made available early to maintain and improve the sports. This money came largely from the sale of hunting and fishing licenses and from special taxes on the industries supported by the sports. The same sources provide most of the funds today. Studies of wildlife species and their environments have produced a great amount of data usable as a basic guide in dealing with other ecological problems.

EARLY WILDLIFE MANAGEMENT consisted almost totally of restricting laws, limiting both the times for hunting and fishing and the amount each person could harvest. Obviously, this was only parceling out an admittedly dwindling resource. The usual next step has been restocking by bringing in replacements from other areas or from hatcheries or farms. Finally, it was recognized that permanent results required improving or restoring the living conditions or habitat for the wildlife species.

Interestingly, similar phases have marked the broader fields of ecology that include human well-being, with recognition now that correcting ills of the total environment is most important.

WILDLIFE MANAGER

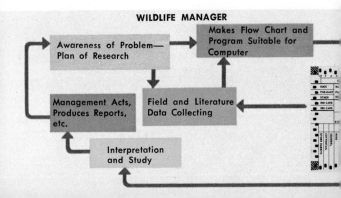

NEEDS OF WILDLIFE are basically the same as for all living things. Each species must have shelter in a suitable habitat, food and water, and an opportunity to reproduce its kind. How these needs are satisfied varies with the species. For this reason, an understanding of the overall environment and of the interaction of its plant and animal components is essential in wildlife management. Each species fits into a food chain that is in turn interwoven with others. Predators are no less important than the prey in the total functioning of the food chains, and neither can exist without the basic food producers—the plants. If any portion of a chain is disturbed, the entire web of which it is a part is affected.

THE MODERN APPROACH to satisfying the needs of wildlife employs computers to store, process, and then analyze data. The kinds of data collected and utilized in making decisions include such things as the effect of land use on the wildlife resource, the computation of total populations by means of capture and recapture of animals, determination of factors that limit the population in an area, and many others.

A comfortable, secure feeling is important for the well-being of wildlife just as it is for people. Factors contributing to this feeling are difficult to detect and measure, but they are a significant part of the ecological picture for wildlife.

COMPUTER SERVICE

SHELTER, or Cover, for wildlife refers to places where animals can retreat and feel safe. In these places they can eat without being disturbed, breed and rear their young, and either play or rest.

SUITABLE SHELTER differs for each species and may vary also with the season. For some kinds of wildlife, it may be necessary to cut or burn trees so that there are open, shrubby areas; for others, perhaps trees must be planted, some allowed to grow taller than others, or dead trees left among the living. Specific needs are learned by observation and research.

In farm lands, patches of crops may be left unmowed and unharvested as wildlife food and cover. Long stretches of uncleaned fence rows can become protected travel lanes rather than dangerous open areas. Brush piles can be located with open "runways" under them.

EDGES and Ecotones are those places where there is a mixture of two or more plant types. Many kinds of wildlife prosper in direct relation to the amount of edges available. Whitetail deer, for example, increased in numbers with the cutting of great forests and the creation of clearings or open areas. Bison, in contrast, thrived in a nearly one-plant environment—the grasslands.

Deep forests do not support as large or varied populations of wildlife as occur along their edges. The animals most benefited by edges are those that move relatively short distances, keeping their food and shelter within close proximity.

The broad, brushy fencerow provides food and cover for wildlife; the clean row has little value to wildlife.

Bobwhites eat mainly seeds, but in summer, insects may constitute as much as 25% of diet.

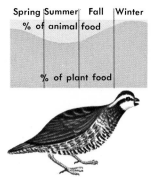

Spring | Summer | Fall | Winter
% of animal food

% of plant food

Spring | Summer | Fall | Winter
% of animal food

% of plant food

Raccoons eat mainly animal food in spring and summer. They depend on plants (about 65%) in the winter months.

FOOD PREFERENCES are distinct for each species of wildlife. Staples are the foods that provide basic energy. In addition, there are foods that a species likes or craves above all others; these are their "sweets" or delicacies. Other kinds of food will be eaten only when the animals are very hungry and can find nothing else. Still others offer little or no nutrition but are eaten because the animal apparently enjoys chewing them. These foods may vary in availability with the season.

FOOD HABITS are learned by research, consisting of close observation, and by analysis of the stomach, crop, and gizzard contents of a large number of animals at different times of the year. The animals are obtained from those harvested during the hunting season, from accidental kills, and from samples taken specifically for these studies. Large animals may be tranquilized and their stomachs flushed.

In addition, droppings or scat may be analyzed, food caches may be discovered and studied, regurgitated pellets of indigestible items are examined, or other methods employed to recognize food needs. The wildlife manager must make certain that the staples are available for the species he expects to maintain. Otherwise the animals may fill their stomachs and still not be properly nourished.

Greater Prairie Chicken cocks on territorial "booming" grounds.

TERRITORY is a specific area that an animal advertises to be its exclusive property and then defends against intruders, particularly of its own species. In species that do establish territories, the total number of animals of that species that can occupy a region is limited by the amount of territorial space available. Not all animals establish territories. Rodents, for example, have only what is called their home range. They allow members of the same species to roam freely over the home range, making no effort to defend it.

DEFENSE OF TERRITORY is greatest during the breeding season. For some birds, the territory may be only slightly larger than the actual nest site. For others it includes additional space in which to collect food. Still others establish territories only for mating, and in some species only the roosting or the feeding areas are defended. The territories of large predators extend over many square miles.

Birds mark their territories by loud calls and by a variety of displays. Other birds of the same species usually respect these announcements by avoiding the territory. In mammals, the territories are commonly marked by scents, which are restrengthened regularly—usually at least once a day. In both birds and mammals, it is the males that exert the claims and drive away other males that may try to gain entry.

Knowing the territorial habits of a species is important in wildlife management. This will determine, for example, how many of the species—say, ducks—can be expected to occupy a nesting area. If nesting boxes or similar aids are to be provided, the territorial habits will guide where they are to be placed. With a knowledge of territorial habits, too, a game manager can make a quick and accurate population census of some species.

LIMITING FACTORS are those important to the survival and productivity of a species. When the limiting factor's availability is either below the critical minimum or in excess of the critical maximum, population growth is halted. These limits of tolerance vary with the species for different kinds of factors, but in each case, it is the limiting factor that must be determined by the wildlife manager and then controlled to increase or improve the population of a wildlife species.

DETERMINING THE LIMITING FACTOR in a complex ecosystem is sometimes difficult, though often it is obvious and easily recognized. Bobwhites, for example, may be extended in their northward range in some areas simply by making certain they are provided with food during times when the land is covered with deep snow. Or bobwhites may be encouraged to move into open country by supplying brush for cover. In both cases it is a matter of correcting the greatest lack to assure an increase in the population. Some other factor then becomes limiting, challenging the game manager to discover and correct it as he works to attain the maximum possible game harvest.

The number of "edges" can be a limiting factor. Note that by increasing the "edges" more coveys of quail exist in same area.

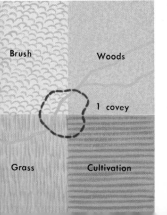

Brush   Woods

1 covey

Grass   Cultivation

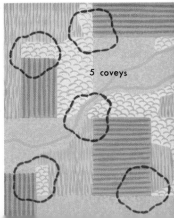

5 coveys

Three or four rabbits (cottontails) can live comfortably on one acre. Each bobwhite, however, needs one or two acres.

A whitetail deer requires about 12 acres—more in some areas, less in others. A cougar or puma needs a thousand acres or more.

POPULATION is the basic commodity in wildlife management, which has as its objective obtaining the maximum harvest for sportsmen without disturbing the breeding stock of the species. This is accomplished by a manipulation of both the wildlife species and its environment or habitat.

In all animals, for example, there is a maximum number, or density, of individuals that can occupy a particular amount of space. This is determined by the ability of the land to supply food, shelter, and whatever special requirements the species has. It is called the carrying capacity of the land. This is best measured at the time of the year when the needs of the species are greatest.

Even if food and cover are available in sufficient amounts, many animals will not exceed a particular population density. The controlling factor in these cases is apparently the animal's need for a particular amount of living space. This maximum density, or saturation point, may be greater or less than the carrying capacity that is determined by the land.

144

MEASURING POPULATIONS consists of taking a census of the number of individuals and, equally important, collecting such data as age, sex, and condition of the population. These factors are important also in determining productivity.

Several censuses at spaced intervals are necessary to learn whether the population is stable, increasing, or decreasing. Comparisons can be made by taking a census on two or more areas at the same time.

DIRECT COUNTS are the most reliable and accurate way of taking a census. Waterfowl may be counted in flocks from the ground, in airplanes, or in boats. Large numbers may be estimated by counting only a portion of the flock and then multiplying to get the total. At nesting time, a census may be made by counting the number of defended territories. Later the number of young can be counted as they feed with their parents. In winter, game birds and animals are commonly censused by tracks in the snow.

INDIRECT CENSUS METHODS give reliable estimates of abundance. These include such techniques as counting the number of shed antlers, the number of coveys of quail that a dog finds in an area, and others.

The Lincoln-Peterson Index is a method that was developed in work with waterfowl but is now used with equal success in determining the populations of other kinds of wildlife, including fish. It was based on the dis-

covery that the number of bands turned in by hunters during a season was a constant percentage of the total number of birds that were banded. This method of censusing works well only in dealing with populations of large size and when there is an accurate tally of one segment of the population obtained by banding, tagging or similar means. Like all indirect census techniques, the measurement gives only relative abundance, not actual population.

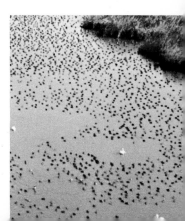

Aerial photography permits accurate count of waterfowl.

BREEDING POTENTIAL is the maximum possible increase in a population under ideal conditions. The breeding potential of a species always exceeds the carrying capacity of the land; food, cover, and other needs are never in adequate supply to support the number of offspring that are produced. Thus only a portion of the young survive in the area. The ultimate population is controlled by environmental factors that prevent growth of the population to its potential.

CALCULATING THE BREEDING POTENTIAL of a species requires a population census. The wildlife manager must also know the minimum breeding age for the species and the number of young produced per year. It is also helpful to know the maximum age at which young can be produced, usual length of life beyond breeding age, and mating habits of the species.

With these facts, the game manager then attempts to discover what factors can be altered to attain the maximum possible productivity. Sometimes a deficiency of food will lower the number of young produced or make the animals more susceptible to diseases, predators, or overharvesting. The basic problem may be poor soil in which food plants grow slowly and which lacks needed minerals and trace elements. By altering such factors, the breeding potential may be more nearly achieved.

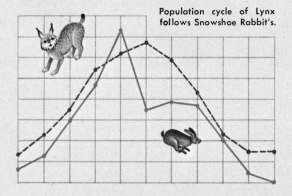

Population cycle of Lynx follows Snowshoe Rabbit's.

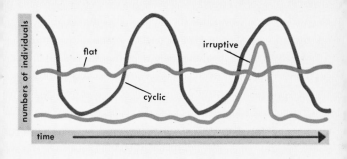

Basic Types of Population Curves

CYCLIC POPULATIONS are those that fluctuate periodically even when environmental conditions remain essentially the same. These are not to be confused with the changes that occur in populations seasonally or with those that occur irregularly and irruptively due to great variations in rainfall, temperature, food, or similar physical or environmental factors.

Snowshoe rabbits provide a classic example of a cyclic population. With no change in its range, the snowshoe rabbit will rise from a low density of only one or two animals per square mile to a high of 1,000 or even 10,000 per square mile. The peaks and lows are usually nine or ten years apart.

Fluctuations in the population of the lynx, which preys on the snowshoe rabbit, follow the same pattern. The peaks and lows lag only slightly those of the rabbit. Similarly, the population of the arctic fox follows the four-year cycle of the lemming.

Among game birds, grouse are distinctly cyclic, reaching highs every nine or ten years. Bobwhites, in contrast, are not typically cyclic, though they do show a tendency toward cycles at the edges of their ranges.

FISH MANAGEMENT principles are fundamentally the same as those for game birds and animals, with the objective to provide fishermen with the maximum catch. Since the 1950's, the greatest effort has been spent in improving fish habitats rather than passing more restrictive laws or concentrating only on stocking programs for "put and take" fishing. It was recognized that, in warm waters, it is virtually impossible to overfish a body of water if the fish have been supplied their basic needs. Surpluses are caught quickly, but as the population of the fish is reduced, the fish become more difficult to catch. Before the lake, pond, or stream can be "fished out," the catchable population is bolstered again by the young of another spawning season.

Well-managed fishing waters do indeed produce as many fish now as in years gone by. The greatest difference is that the fish must be divided among many more fishermen. Pollution and lowered water levels are the greatest threats to fishing today. These changes make the waters no longer suitable habitats.

Removal of large "wolf" bass past reproductive capacity is good conservation, eliminating non-productive consumers.

In ponds and small lakes, a "bloom" of algae is often killed by spreading copper sulphate in the water, as shown here.

duck nests
in farm pond

FARM PONDS have provided an excellent opportunity to manipulate an environment and its population. Most ponds, which average an acre in size, were first created to have a reserve water supply for livestock. Then it came in vogue to stock them with fish, mainly to enjoy the sport to be had in fishing for them.

STOCKING of farm ponds is generally with largemouth bass, which are predator fish, and with bluegills, which eat insects and other small aquatic animals and then become food for the bass. In the early days, the fishing generally became poor after about two years. Studies of the fish population revealed why.

In a typical stocking, 10 bluegills are placed in the pond for every 1 bass. The bluegills multiply much more rapidly, and soon there is an overpopulation. The bluegills are stunted, with oversized heads and paper-thin bodies. An acre of water, like land, can support only a limited amount of livestock.

Pond owners learned that they had to catch many more bluegills than bass to keep the population of their ponds healthy. Even with greater fishing pressure, the bluegills eventually gain, and so ponds are now built with drains and catch basins so that the population balance can be restored periodically.

Ponds supplied with water from streams fill too rapidly with silt. The stream also brings in a variety of unwanted fish that take food and living space from the bass and bluegills. It is best for the water to come by seepage from a watershed.

Weeds soon grow around the edges of a pond if the water is shallow. Uncontrolled, they creep outward from the shore toward the center. They become a refuge for fish—usually the bluegills that can then no longer be caught easily by hook and line or by the predatory bass. A pond should be three feet deep or deeper at its shores.

Fertilizers must be used wisely. If too much is added, the algae in the pond "blooms," consumes the oxygen, and kills the fish (p. 118). If the algae do not grow at all, however, the plankton-feeding aquatic life goes hungry, in turn affecting all animals up the food pyramid to the bass, which are the top predators in the pond.

LAKES can be improved for fishing most immediately and most permanently by making certain their watersheds are not feeding silt or chemical pollutants into the water. "Cleaned up" lakes will in most instances produce fish to the limit of their carrying capacity without assistance. Some of the efforts of other sorts to improve conditions for fish and fishing are of questionable value as indicated below.

PROVIDING BRUSH SHELTERS in which young fish can hide is a common practice that should be avoided in most situations. The fish using these "hideouts" are mainly panfish that are likely to become too numerous anyhow. It is best to keep the waters open so that the predators can make meals of the panfish more easily. Further, the brush makes fishing more difficult, because lines and lures become easily snagged, hampering harvests by sport fishing.

Weed beds generally become problems, too. Fishermen cannot fish in shallow, weedy tangles, and the fish that use the weeds as a sanctuary are again mostly the smaller, too plentiful panfish. Only in exceptional cases, as with brush shelters, is it good fish management to create weed beds. Controlling weeds is a major problem.

In some instances a limiting factor may be identified, and if it can be corrected, the fish population may increase dramatically. In lakes used as reservoirs to supply water for generators, the insect population is usually low because the water is drawn down periodically and the immature insects cannot develop. Panfish that depend on insects for food do not prosper in these lakes, and so the game fish predators that eat the panfish are also limited in numbers. Introducing plankton-feeding fish that can get their meals in the open water and are unaffected by the lowering of the water level creates an entirely new kind of food chain. The plankton-feeding fish become food for the game fish, which then become plentiful.

Previously open body of water is now covered with "weeds."

Fish "ladders" help migrating salmon past impoundments.

STREAMS may flow fast or slow, and their waters may be cold or warm. Each type of stream has a characteristic population of fish and other aquatic life. Broad, slow-moving streams are much like lakes, the water in their pools scarcely moving. Life in these streams must have low oxygen needs. Fast-moving streams have a high oxygen content. They are the habitat of trout, darters, and other kinds of life with high oxygen needs.

But no matter what its type or its basic, native population, a stream's productiveness depends on the character of the water fed into it from its watershed. Polluted streams can support only limited amounts of life—and eventually almost none.

POOR FISHING in streams is an ill generated from the land through which the stream flows. Many years may be required to restore a stream to productiveness, but the effort not only improves the fishing but also the land. When fast runoffs of water are stopped by converting the watershed again into a green sponge of forests and other vegetation, destructive floods are also eliminated.

Stream improvements for fish are rarely needed except where man has disrupted the watershed. Dams, current deflectors, shelters, artificial spawning sites —all of these improvement devices are not wise investments and will do little or no good until the basic problem of restoring the habitat conditions on the land is accomplished. Good fishing develops naturally when a stream's watershed is healthy.

151

Brown Pelican

Whooping Crane

ENDANGERED SPECIES

The extinction of species, both plants and animals, has occurred throughout the history of life on earth. It is estimated that for every species alive today as many as a hundred have become extinct. The rocks of the earth are filled with the fossils of plants and animals that no longer exist.

Most natural extinctions take place gradually over many thousands of years in the course of evolution. Ordinarily they come about as a result of genetic differences that eventually produces a new species. In the unrelenting competition for food and space, the new species occupies the niche of an older species and eventually eliminates it.

Other extinctions have occurred with dramatic suddenness when there have been abrupt, major changes in the environment. Most quickly affected in these cases were those plants and animals that had become highly specialized. They had literally traveled an evolutionary "blind alley" from which they could not retreat and were unable to adapt to the new conditions. Species with a wide range of tolerance to living conditions and foods are successful over the longest periods of time.

Geological extinctions are measured on such a vast scale of time that their rate is truly not alarming. The extinctions that have occurred due to man's changes in the world environment have come about with much greater speed. They are occurring now at the rate of about two per year. Most sobering, they are happening because of changes in ecosystems of which man himself depends for his survival. Unwittingly, man has placed himself on the list of endangered species.

Ecologists have still another concern about the losses of plants and animals. With each extinction, unique combinations of genetic traits are lost forever. As a meaningful example, some of the wild plants from which cultivated crops have been developed may be more tolerant to climatic changes and more resistant to diseases than are the new, improved, high-yield varieties. The wild plants must be preserved as a "bank" of genes that can be turned to in case the improved varieties somehow meet disaster.

Assignment of specific values to losses of plants and animals is not always easy, but the interaction of all living things is so complexly interwoven that no loss can be considered insignificant. Directly or indirectly, each loss affects many other forms of life and is generally a symptom of an environmental ill. There is no way of determining which plant or animal might have future value or for what reason. There is no valid reason for permitting any extinction.

Komodo Dragon, a giant lizard 10 feet long and weighing 250 pounds, is an endangered species on its Indonesian island.

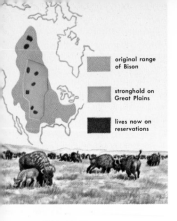

original range of Bison

stronghold on Great Plains

lives now on reservations

SLAUGHTER has victimized animals prized as trophies or for their hides, feathers or meat.

Flocks of passenger pigeons in America in the early 1800's were estimated to contain as many as 50 million birds. Before the end of the century, market hunters had reduced their numbers to fewer birds than could survive. The last passenger pigeon died in the Cincinnati Zoo in 1914. The heath hen, once common, was also hunted to extinction.

Annihilation of animals by slaughter has not been confined to heavily populated areas. The dodo, turkey-sized flightless bird that lived on Mauritius Island, some 600 miles east of Madagascar, was unknown to man until 1598 when it was discovered there by Dutch explorers. In less than a hundred years, the dodo was destroyed.

Tigers, whales, wolves, Galapagos tortoises—these are among the many animals endangered by unnecessary slaughter and harvesting.

Bison's natural range has been decreased to limited areas as shown on map.

REDUCED SPACE is another factor contributing to the disappearance or endangering of species, true especially of large animals.

The American bison, or buffalo, seemed doomed but was saved. Its population of about 60 million was reduced to less than a thousand in about 50 years by hunters. These few animals were moved onto federal lands where the herds are managed and harvested scientifically to control their numbers.

Plants have been affected, too. Some authorities say that as many as 10 percent of the total population of flowering plants are endangered. The tall, spectacular saguaro cacti of southwestern America are being endangered by the steady settlement and reduction of the amount of desert area. Additional factors make the situation more complex. Stockmen killed predators they believed were a threat to their livestock. As a result, rodents increased in numbers and now eat nearly all the seed produced by the cacti. Natural seeding has thus been greatly reduced. Cattle trample young cacti that do get started and also eat or beat down the desert shrubs that provide the shade needed by the cacti in early growth stages. A cactus seedling ten years old may still be only a few inches tall.

OISONS have endangered particularly the predators due to biological magnification (p. 107). Peregrine falcons, pelicans, and other birds of prey are among the species suffering from chemical poisoning. A number of species of fish exist now in sparse numbers because waters have been made unlivable. Streams, lakes, bays, and estuaries have become open sewers and catchments for drainage systems.

Poisons become a special hazard when animals are migrating or are aggregated for breeding. Application of pesticides on crops by fast-moving ground vehicles and by low-flying airplanes douses the animals and their perches. Drift of fine agricultural chemical sprays has been responsible for a number of bird kills.

Governments now set limits on allowable pesticide and herbicide residues on agricultural products grown for consumption by people and livestock. The accumulation of chemicals and heavy metals in foods is now monitored to keep concentrations at acceptable levels.

DISEASES also take their toll. American elms and chestnuts are both endangered plants because of introduced fungus diseases that have spread through their populations rapidly.

The fungus causing the disease in chestnuts arrived in ports in the Northeast early in the 1900's. Once abundant, the chestnuts are now reduced to a few short-lived sprouts from old trunks and roots.

The elm fungus, spread by elm bark beetles, also arrived in Northeast ports sometime during World War I. Newly diseased trees are located by aerial infrared photography (p. 130) and are removed in efforts to prevent the spread of the disease. But despite these efforts, the elms are rapidly disappearing throughout the United States.

Animals are most likely to become diseased when they are crowded at water, feeding, and breeding locations. Aggregations of this sort are most common among animals that migrate. If their population is already low or weakened due to other factors, a rapidly spreading disease can be devastating.

Diseased tree detected by aerial infrared photograph is marked for removal in fight against the Dutch elm disease.

Migration paths of the Green Turtle are tracked by fastening a radio transmitter to the turtle's back and releasing it in the sea.

SPARED SPECIES are fortunately increasing in numbers as man learns more about correcting mistakes made in the past.

Some plants have escaped extinction as a result of having been taken into cultivation. Among these are redwoods, Montery cypresses, and cedars of Lebanon. The ginkgo, or maidenhair tree, is the sole living representative of a group of plants with an evolutionary history of more than 200 million years. Once this tree was naturally distributed over a wide region of the world, but in the climatic changes that occurred, it survived only in a limited region of eastern Asia. There, in historic times, Buddhists began planting the trees around monasteries. Easily propagated, the ginko is now grown as an ornamental in temperate regions throughout the world and is no longer rare or endangered.

Once the green turtle was compared in its abundance in the warm seas of the American tropics and subtropics to the vast herds of bison on the Great Plains. Turtle fishermen from many countries hunted the turtles for their meat and robbed the nests on the beaches to take the eggs. Wild dogs and other animals also ate the eggs. Now there is an international protective association working to save the turtle from extinction.

American alligators have been spared, hopefully, by protection in parks and by laws that prevent the killing of alligators for hides or taking young animals for pets. Crocodiles exist in North America on government lands.

Whooping cranes have apparently been spared by the creation of refuges on their winter and summer ranges and by the successful hatching of the eggs in captivity. Many of the best-known animals of Africa can be found now only in the sanctuary of the many parks and refuges being established there.